MANUFACTURING SUBURBS

MANUFACTURING SUBURBS

BUILDING WORK AND HOME

ON THE

METROPOLITAN FRINGE

EDITED BY
Robert Lewis

Temple University Press
PHILADELPHIA

Temple University Press
1601 North Broad Street
Philadelphia PA 19122
www.temple.edu/tempress

⊗ The paper used in this publication meets the requirements of the American National
Standard for Information Sciences—Permanence of Paper for Printed Library Materials,
ANSI Z39.48-1992

Library of Congress Cataloging-in-Publication Data

Manufacturing suburbs : building work and home on the metropolitan fringe / edited by
   Robert Lewis.
      p. cm.
   Includes some original papers commissioned for this collection and some [revised
from those] previously published in issues of the *Journal of historical geography* and
the *Geographical review*.
   Includes bibliographical references and index.
   ISBN 1-59213-085-2 (cloth : alk. paper) — ISBN 1-59213-086-0 (pbk. : alk. paper)
   1. Suburbs—United States—History. 2. Suburbs—Canada—History. 3. Manu-
facturing industries—United States—History. 4. Manufacturing industries—Canada—
History. 5. Working class—United States—History. 6. Working class—Canada—
History. 7. Urbanization—United States—History. 8. Urbanization—Canada—
History. I. Title: Work and home on the metropolitan fringe. II. Lewis, Robert D.,
1954–

HT352.U5 M36 2004
307.76'0973–dc22   2004041236

2 4 6 8 9 7 5 3 1

Contents

Preface

This book, an examination of industrial suburbanization and the making of industrial suburbs in Canada and the United States between 1850 and 1950, has three aims. The first is to assemble in one collection, for the first time, a selection of recent research on industrial suburbanization and industrial suburbs. Building on historiographic and theoretical overviews, case studies of various metropolitan areas in Canada (Toronto and Montreal) and the United States (Baltimore, Pittsburgh, Chicago, Detroit, Los Angeles, and San Francisco) highlight the place-based nature of the factory districts encircling the central city.

The second is to help refocus research on the suburbs. Until recently, most work favored middle-class residential suburbs. Along with other work that has been published in recent years, this book looks at industrial and working-class suburbs. The third aim is to show the significance of the dispersion of factories to the urban fringe for the making of metropolitan districts before the post–World War II bout of suburbanization. The book argues that manufacturing decentralization was a vital component of metropolitan development from as early as the mid-nineteenth century.

Some chapters are original papers specifically commissioned for the collection (Chapters 1, 4, 8, 10, and 11). The rest have appeared elsewhere. Chapters 2, 5, 6, 7, and 9 were originally published in a special issue of the *Journal of Historical Geography*, volume 27 (2001). Chapter 3, which is arguably the earliest academic examination of factory districts, and the model and forerunner of most other research on metropolitan industrial districts, was first published in the *Geographical Review*, volume 26 (1979).

In one way or another many people have contributed to the making of this book. Various colleagues have provided intellectual support, but I would like to give special thanks to Sherry Olson, Richard Harris, Gunter Gad, Ted Muller, and Dick Walker. At the University of Toronto, Joe Desloges, chair of the Geography Department, provided financial and administrative support, while Byron Moldofsky, Jane Davie, and Mark Fram helped me through the nightmare of producing illustrations in the electronic era. A grant from the Social Science and Humanities Research Council of Canada and a fellowship from the Connaught Committee of the University of Toronto gave me the time to write part of the book and organize the rest. I have benefited from the wisdom of Peter Wissoker, a senior acquisitions editor at Temple University Press, who has guided the book through some treacherous shoals. Finally, the lights of my life, Lisa, Yonah, and Lev, have provided me with love and support, and I am deeply grateful.

MANUFACTURING SUBURBS

1

Industry and the Suburbs

ROBERT LEWIS

The industrial origins of metropolitan Chicago's Pullman and South Chicago districts date back to the late 1860s, when Colonel James Brown and other prominent Chicagoans purchased a large tract of land and secured a charter for the Calumet and Chicago Canal and Dock Company. The building of the harbor and rail lines, the opening and grading of streets, and the provision of services helped attract the area's first industries, among them Northwestern Fertilizing (1867) and James Brown Iron and Steel (1875). The most singular events, however, the ones that would transform the somewhat sleepy suburban districts south of the city limits, were the building of the North Chicago Rolling Mills and the Pullman Palace Car Works in 1880. By 1882 the former consisted of four blast furnaces, a Bessemer steel department, and a steel rail mill, thus setting the foundation for the development of one of America's largest steel complexes. Pullman became the nation's largest railroad car producer and most notorious representative of the corporate industrial suburb. By the time the two districts, as well as an assortment of other suburban industrial communities on the south side, were annexed to Chicago in 1889, the foundation for tremendous economic and residential growth was in place: industry hugged the lake shore, the Calumet River, and the railroad lines, while working-class housing was sandwiched between industry, water, and rail.[1]

Whenever the subject of pre-1950 industrial suburbs, such as South Chicago and Pullman, comes up in discussion, the almost inevitable response from urban history scholars is, "everyone knows about those." Regardless of their core research interests, scholars have a strong sense that industrial suburbs have littered the metropolitan fringe for a long time. Those studying the late eighteenth and early nineteenth centuries point to the old textile suburbs of Philadelphia, the fringe districts where noxious, land-intensive or material-oriented enterprises such as tanneries, shipyards, and brick-making yards clustered, and to the nearby water-powered industrial villages and towns.[2] Those with research interests in the nineteenth and early twentieth centuries refer to the "satellite cities" of Chicago, Cincinnati, and St. Louis, and, of course, most commonly to the Chicago suburb of Pullman.[3] Scholars working on particular cities point to the industrial suburbs clinging to the urban fringe: interwar metropolitan Detroit was composed of Dearborn, Ecorse, Highland Park, Pontiac, River Rouge, and Wyandotte; Buffalo was surrounded by Lackawanna, North Tonawanda, and Tonawanda; early twentieth-century St. Louis was encircled by East St. Louis, National City, Granite City, and Alton; and Milwaukee was belted by Cudahy, South Milwaukee, Waukesha, and West Allis.[4]

Although it appears that we know a great deal about industrial suburbs before World War II, the places on the metropolitan fringe where men, women, and children worked in factories, workshops, mills, railroad shops, and warehouses have, until recently, been taken for granted or largely ignored. Despite the identification of a large number of manufacturing areas on the metropolitan fringe, few historical studies have focused on the industrial suburb or industrial decentralization. What we do know about the character of these nineteenth- and early twentieth-century places and the processes producing them comes from a handful of specialized studies or the occasional paragraph in an urban and industrial history. The ringing of the central city by industrial suburbs as a primary research focus has been of secondary interest to urban scholars.

One reason we know relatively little about the industrial suburb before the 1940s is that scholars working on the post–World War II metropolis have downplayed the few studies that have been done. Instead they have emphasized what they consider the tremendous rate and unique character of manufacturing and retailing decentralization after 1950. In this view, economic decentralization before the 1920s was negligible and amounted to only a trickle in the interwar period. For them, the real impact of manufacturing suburbanization and the development of working-class industrial suburbs lies in the postwar period.[5] We actually know less about the prewar industrial suburbs and industrial suburbanization than we think we know.

Another reason is that the tremendous advances in our knowledge of middle-class suburbs before 1950 have overshadowed research on other aspects of the metropolitan fringe. The place of industrial suburbs within the broader dynamics of urban growth has been lost in the historiography of the middle-class suburb. Scholars have paid much greater attention to the reasons for the development of the residential suburb and the processes driving residential suburbanization. This is nowhere more apparent than in the growing number of (mainly middle- and upper-class) residential suburban histories that have appeared since the publication of Sam Bass Warner's *Streetcar Suburbs* forty years ago. Since then, urban scholars have examined many aspects of the middle-class residential enclaves around the expanding metropolitan fringe after 1850.[6]

While this residential literature is extensive and has taken many different tacks, it is possible to identify several main themes. One major focus of research has been to identify the mechanisms of the suburban land development process, from the local alliances that converted farm to urban land to the origins and meaning of the new designs that developers constructed on the metropolitan fringe. Scholars have dissected the process by which the middle class sought an idyllic rural/urban world where they could escape the central city's congestion, noise, and working class, while having access to both central-city jobs and a suburban lifestyle. This, in turn, is related to studies tying the creation of this leafy suburban world to the formation of an increasingly segregated metropolitan community, where residential opportunities based on gender, class, and race became ever more exclusionary. Finally, many researchers have pointed to the importance of the commuter railroad, the streetcar, and the automobile in opening up the metropolitan fringe to developers and the middle class. In sum, our knowledge of the activities of the urban fringe has been greatly enriched by this work on the crabgrass frontiers of the bourgeois utopias and the streetcar suburbs.

The History of Research on the Industrial Suburb

Although middle-class suburbs have been the focus of research, there is a long, though relatively meager, history of work on industrial suburbs and industrial suburbanization. One aspect of this work, dating from the 1930s, focuses on the compilation of the metropolitan geography of manufacturing employment. Using census-defined city-suburban boundaries, Glenn McLaughlin and Daniel Creamer pointed to the large number of factory workers in the suburbs and, in the process, laid the basis for the

appreciation of the larger metropolitan picture of industrial suburbanization.[7] Prompted by the noticeable bleeding of central-city jobs to the metropolitan fringe and the changing balance of central and suburban employment, these early studies demonstrated the degree to which these changes
were taking place. In the early postwar period writers such as Coleman
Woodbury and Frank Cliffe, building on the work conducted during the
1930s, pointed to the continued significance of industrial decentralization
during and after the war.[8]

Another set of work examined individual industrial suburbs or the
industrial suburbs of a few cities, using other types of evidence along with
aggregate census data. The classic study is Graeme Taylor's 1915 description of the satellite cities surrounding Chicago, St. Louis, and Cincinnati.
While polemical in style and selective in use of evidence, Taylor's book
was nevertheless the first attempt to examine the character of the industrial fringe. What Taylor conclusively revealed, a generation before
McLaughlin's and Creamer's more aggregated work, was the importance
of manufacturing employment on the metropolitan fringe since the 1880s.
At the same time, the Pittsburgh Survey's investigation of the metropolitan area's mill towns, and the Regional Plan of New York's examination
of New York City's, established the importance of decentralization at the
turn of the century.[9]

This early work was followed by a series of case studies and suburban
histories. The existence of industrial activity and the concomitant development of working-class residential areas in the metropolitan suburbs that
Taylor described was noted by Harlan Douglass in the mid 1920s.[10] This
was followed in the postwar period by a spate of studies that discussed the
making of industrial suburbs before 1950. Foremost among these were
investigations of mid-nineteenth-century Baltimore's factory districts, the
Chicago suburbs of Pullman, Cicero, and Skokie, the automobile suburbs
of twentieth-century Detroit, and New York satellite towns such as
Paterson-Passaic.[11] In addition to census data, these authors used an array
of evidence, including insurance atlases, industrial directories, and firm histories, to explain the motivations of firms moving to the metropolitan
fringe and to illustrate the relationship of industry to a growing working-
class suburban population. This small number of early postwar studies, for
the most part undertaken by geographers and historians and focusing on
industrial activity as much as on social aspects of industrial suburbs, stood
in direct contrast to the mostly sociological studies of working-class suburbs, which directed their attention to residential suburbs and were more

interested in the social ramifications of suburban life than in the industrial aspects of the suburbs.[12]

Finally, in the 1970s there appeared studies of the political economy of metropolitan growth, with particular stress on suburban development.[13] The intent of writers such as David Gordon and Richard Walker was not so much to describe and explain the lived-in world of industrial and working-class suburbs as it was to explain industrial suburbanization. This was achieved by working at a highly abstract Marxian level, in which concepts such as mode of production and capital-labor conflict shaped the interpretation of metropolitan development. Although each writer had his or her own particular position, the general argument was that the industrial metropolis was the creature of the capitalist mode of production, and that metropolitan spatial form was contingent upon capitalist accumulation. Changes to urban form and industrial and residential suburbanization were structured by class conflict and emerged as a solution to capitalism's internal contradictions. The corporate city, for example, was defined by the decentralization of industrial activity as capitalists sought factory sites away from worker unrest in the central city.

This analysis of metropolitan form paved the way for the development of a much-stripped-down version of Marxian theory, with the attempt by Allen Scott to merge political economy with transactional theory.[14] Using a basic division between large, stand-alone, vertically integrated firms (as in the chemical and steel industries) and small, face-to-face, horizontally integrated firms (as in the clothing and jewelry industries), Scott constructed a metropolitan fault line: large firms sought out the urban fringe, while small firms sought the face-to-face milieu of the city center. As is described in greater detail in Chapter 2, Scott, Walker, and the others working in the political economy vein sought to replace the traditional factors of locational analysis with a more incisive and class-based theory.

This long lineage of research has generated a body of work on the industrial suburb. The work of McLaughlin and others, even though couched at a highly aggregated level, demonstrates both the growing number of suburban manufacturing jobs and the changes to metropolitan America's geography of employment, both before and after World War II. The less census dependent, more qualitative studies of Muller, Groves, Taylor, Buder, and Kenyon took these points a step further by bringing out the strategies of the firms seeking suburban locations and emphasizing the tangible workplaces built beyond the city line from as early as the first half of the nineteenth century. The political economy of the metropolitan growth perspective placed industrial suburbanization and the making of industrial suburbs within a broader context, by extending this earlier analysis of individual firm behavior with a theoretical focus on the class character of urban-industrial society. While not substantial compared to the comprehensive body of suburban social geographies and histories, these pre-1980s studies of the industrial suburb and industrial suburbanization have nevertheless proved to be an extremely valuable guide to the long and changing history of the metropolitan fringe. Regardless of the theoretical position, methodological approach, and choice of sources, these writers identified a powerful and distinct process of industrial suburbanization and the development of a ring of industrial suburbs on the metropolitan fringe.

Building on this literature, a new round of work following two major lines of thought has added to our knowledge of the metropolitan suburbs in the past twenty years. One line consists of studies that examine the working-class and African American suburban residential experience. The narrative of the middle-class suburb that has dominated postwar research on the expanding metropolitan fringe before 1950 has been complemented by examination of the lifeways, housing strategies, political choices, and work experience of white and black workers, both native-born and immigrant. From Toronto and Hamilton's owner-built suburbs to the black towns and working-class suburbs of Los Angeles, Philadelphia, Detroit, Knoxville, Birmingham, and Montreal, the notion that the suburbs are largely the home of middle-class commuters has been successfully challenged. The focus on the social features and experience of the working-class and industrial suburb has greatly advanced our understanding of the suburban mosaic.[15]

The second line of recent scholarship focuses first and foremost on suburban industrial activities and employment histories before 1950. Writers have drawn detailed studies of specific industrial suburbs, satellite cities, and industrial towns ringing the metropolitan central core. This has been done primarily by blending archival methods and extensive bodies of evidence with various theories about firm behavior and urban development.[16] While each researcher has pointed to the unique features of his or her subject, the focus has remained on some aspect of the origins and development of industry on the metropolitan fringe. From Montreal's industrial suburb of Maisonneuve, where a French Canadian bourgeoisie built factory spaces and working-class housing, to the oil suburbs of Los Angeles and the factories and blue-collar homes of suburban New York, the importance of the industrial suburb has been documented.

Themes and Issues of the Volume

The result, as further witnessed by the collection of papers in this volume, is that scholars over the past twenty years have developed a large and coherent body of knowledge about the diverse social and economic character of the ever-changing North American urban fringe in the nineteenth century and the first half of the twentieth.[17] Building on earlier studies, this recent research has illuminated various features of the industrial suburb, notably the geography and scale of industrial suburbanization since the 1850s, the relevance and character of city-suburban boundaries, and the reasons why firms moved to the suburbs.

This concern with the relationship between the rise of the industrial suburb and capitalist industrialization has pointed to the definite increase, in both absolute and relative terms, in the number of metropolitan manufacturing jobs found in the suburbs. The rise of the industrial suburb after the middle of the nineteenth century is clear. The number of suburban jobs increased dramatically after 1850, and especially after 1880. In sixteen of America's largest metropolitan areas, the number of suburban jobs rose from 446,000 in 1879 to more than 1.6 million in the interwar period and almost 2.5 million by 1954. Suburban employment growth outpaced central-city growth in this period. In the second half of the nineteenth century, suburban manufacturing employment made up about a third of all metropolitan employment. As the number of suburban jobs grew faster than those in the central city after World War I, the suburban share reached almost half of all metropolitan jobs by 1954.[18]

Factories, mills, and warehouses clustered on the urban periphery from an early date. In some cases, industry huddled in suburbs where few other activities were to be found. In others, factories intermingled with workers' housing. In city after city, the surrounding fringe of the expanding metropolis was home not just to middle-class stockbrokers and clerks but also to machinists and unskilled steel workers. The historical geography of the suburbs was as much about the factory and the homes of workers as it was about the middle and upper classes. Of course, the character of industrial suburbanization differed from city to city: Pittsburgh was not Montreal; San Francisco was not Toronto. But regardless of the particularities of economic structure, ethnic composition, local growth machines, and transportation politics, industrial suburbanization patterns in Canadian and American cities had much in common.

The chapters in this volume bring out several important and common aspects of industrial suburbs and industrial suburbanization. The first has to do with the relationship between the development of the suburb and urban building rhythms. New fringe development and industrial landscapes were created in waves of industrialization and building construction. After the 1890s depression, for example, investment, industrial growth, and city building were focused on San Francisco's metropolitan fringe (South San Francisco, Oakland, and Contra Costa) rather than on the city itself. In favoring the metropolitan periphery, industrial growth enabled the formation of an elaborate multinodal industrial geography encompassing a large territory that hugged the bay. Similarly, industrial development in Pittsburgh, Toronto, Detroit, Chicago, and Montreal created districts farther out from the city core in each burst of growth.[19]

In their search for new location sites, entrepreneurs of all stripes channeled capital investment away from existing districts, with their poor infrastructures, volatile labor forces, inadequate building structures, and congested housing districts, to greenfield sites on the metropolitan fringe. The notoriously lumpy, volatile, and imitative character of capital investment ensured that large amounts of capital would flow into newly established factory sites. In each building cycle, localities laid down fixed capital in plant and constructed the prerequisites for a viable industrial environment: infrastructures (sewers, electricity lines), transportation (streets, railroad lines, canals), and elements of social reproduction (housing, stores, churches).

Linked to the development of industrial suburbs was the sorting of the metropolitan area's social geography. Especially important here, as Richard Harris points out in Chapter 11, is the home-work linkage. The relationship between where people worked and where they lived is central to a better understanding of the suburbanization process and the social character of the suburbs. Factory districts themselves often became the home of workers employed in the suburb's factories, workshops, railroad yards, and mills. The steel mills, foundries, and metalworking shops of Montreal's West End pulled large numbers of skilled and semiskilled metal workers (such as puddlers, molders, and fitters) and an even larger body of unskilled laborers into the district. Pittsburgh's steel towns housed a largely captive labor force, while Detroit's automotive suburbs were home to a distinctive workforce. The rise of Chicago's Union Stock Yards and the associated meat-working plants from the 1860s led to the building of the Back of the Yards, a heavily populated multiethnic working-class area that housed the vast majority of the Packingtown's workers.

Factory districts also shaped the residential geography of the middle and upper classes. The development of new and the continued existence of older fringe industrial districts, with their noisy, dirty factories, waterways, and railroad tracks, their cheap and densely packed housing, and their working-class and immigrant populations, forced white-collar professionals and the bourgeoisie to settle elsewhere. In the second half of the nineteenth century, class, racial, and income-based geographical sectors became increasingly important social dividers in places such as San Francisco, where the wealthy fled the older districts and settled on Nob Hill and places farther west. Once in place, these new residential patterns shaped the metropolitan social geography for generations to come.

Another issue examined in this volume is the importance of the organization of production for the development of the industrial suburb. Eschewing the polarized descriptions of industrial organization, with their emphasis on the difference between large, vertically integrated, multiunit, dynamic corporations and small, stagnant, labor-intensive firms, this study builds on an alternative model of industrial organization. Industries functioned in quite different ways and had a diversity of social, technological, market, and material properties. This produced an array of production strategies, which in turn produced a range of workplaces and organizational methods.[20] There were many industrial capitalist trajectories, from mid-nineteenth-century Baltimore, with its coexistence of commercial and industrial capitalist organizational forms, to the extremely varied industrial structures that developed in Montreal, San Francisco, Los Angeles, Pittsburgh, and Toronto.

The diverse production pathways that characterize metropolitan development produced quite different locational requirements. In each wave of industrial growth, the parameters of what constituted acceptable sites for production were restructured in response to the development of new production, process, and transportation technologies, the changing character of the labor force and inter-firm relations, and the growing scale of production and markets. Industrial restructuring created the need for new sites, many of them in the industrial suburbs and towns on the expanding metropolitan fringe before World War II. These areas provided large lots of cheap land where firms could implement new technologies and build new factories away from the congested central city, with its large and increasingly hostile labor force. Although large integrated firms typically initiated suburbanization, the development of suburban agglomeration economies attracted firms of all sizes from a range of industries and produced a diversity of productive strategies on the metropolitan fringe.

The development of production trajectories and production sites on the urban fringe was tightly bound up with the emergence of a territorially bound set of functions that traversed the entire metropolitan district. Business and inter-firm networks operated at the metropolitan scale. In Pittsburgh, for example, steel manufacture depended on the ability of all types of business interests, ranging from financial institutions in central Pittsburgh to owners of coke ovens in the city's rural hinterlands, to coherently organize their interests. Each burst of industrial growth in Toronto and Chicago relied on the ability of property interests to collate land into a manageable commodity, subdivide the new commodity in a manner appropriate to industry, and connect the new sites with transportation systems and working-class housing. The development of locational assets in one part of the metropolis was connected to what went on in other parts. The functional linking of the strategies of urban elites and entrepreneurs was critical to the development of the industrial suburb.

Many of the contributors to this volume examine the impact on factory district formation and suburban development of events and processes occurring at various spatial scales. Metropolitan development and industrial suburban growth were shaped not only by local manufacturing entrepreneurs, financiers, property developers, politicians, and transportation interests. Also important were processes operating at the regional, national, and international levels. Baltimore's mid-nineteenth-century industrial geography, for example, was shaped by capitalist industrialization operating at other scales, including nonlocal technological change and labor transformations, expanding regional and national markets, the changing character of global (mostly Atlantic-based) networks, and the city's place in the nation's developing urban hierarchy.

Similarly, the rapid growth of an urban-industrial complex in Chicago, Pittsburgh, and San Francisco was deeply rooted in the exploitation of natural resources, particularly grain, lumber, meat, and minerals. From the California goldmines to the midwestern wheat fields and forests to southwestern Pennsylvania's clay, gas, and coal deposits, the distribution and exploitation of natural resources contributed to the character of metropolitan industrial development. As these far-flung networks of capital, technology, information, resources, and labor coalesced at the local scale and functioned within waves of investment, industry took shape in various locations—the central city, the adjacent suburban districts, and the more distant satellite towns.[21]

The chapters in this volume also touch upon the relevance of the boundaries separating city from suburb. There is no disputing the political and cultural importance of these boundaries. Suburban life, especially before the development of the metropolitan or regional administrative bodies that oversee and coordinate many elements of urban life, often meant better access to educational and social facilities, a greater range of utilities and housing, and a more spacious and cleaner environment. It also allowed blue-collar homeowners to skirt what they considered onerous government regulation and expensive urban real estate. For the rest of the working class, living in the suburbs meant easy access to suburban and, if connected by streetcar, central-city jobs. It certainly provided the potential, regardless of class, ethnic, and racial background, to construct a suburban identity, one predicated on city-suburb differences.[22]

With few exceptions, however, most of these supposed suburban assets could be found within the city, especially in the newly developed areas outside the preexisting built-up core. This can be seen in Atherton Park, where an "alternating" process of suburbanization occurred within Flint's city boundaries. New neighborhoods, housing people from all backgrounds, containing various kinds of industry, and differing little from suburban ones, could be and were constructed within the city boundaries. Ironically, to complicate matters, many factory and working-class developments within city boundaries were originally suburban. This was especially the case during the great annexation wave of 1850–1918, when large chunks of the metropolitan area's outer ring were annexed by the central city. Not only did these annexed districts increase the population and number of jobs in the city, they also provided room for cities to grow and expand.[23]

Years of political wrangling notwithstanding, these areas were defined as suburban one day, urban the next. What this meant for the new industrial suburban dweller or firm varied, of course. For many firms and people, the difference must have been small. Certainly, in tandem with what was occurring in the suburbs ringing the city, there was great variation in the quality and volume of goods and services in this new city landscape. It can be argued that, with the exception of control over municipal government—admittedly no small feature, but nevertheless only one aspect of daily life, and for most residents probably only a small one—there was little to separate the newly constructed parts of the city from the new suburb.

If newness is one criterion for defining the suburb, then the city-suburb divide, once again, is problematic. There can be little doubt that there is a gradient leading out from the city's central core, where the areas tend to be younger the farther they are from the center. This, of course, was the picture that Ernest Burgess drew with his concentric ring model in the 1920s, which has been re-created in a various guises by many other urban scholars, among them Leo Schnore and Sam Bass Warner.[24] Although many writers tend to emphasize the newness of the suburb relative to the city, there is, as various chapters here demonstrate, little support for this view. Like the social gradient, the building history gradient has little basis in tangible metropolitan form.

As the contributors to this book also stress, the conversion of the urban fringe into industrial and working-class suburbs was not a process imposed on an empty slate. Fringe factory districts were frequently molded out of or made to conform to an area's natural habitat. Industrial, real estate, and political elites were extremely adroit at shaping the natural environment to their needs. For example, elements of Chicago's local bourgeoisie, with the support of state and national interests, were instrumental in making the Calumet and Chicago Rivers major industrial channels. Similarly, Pittsburgh's elite fashioned a particular steel-producing geography out of the valleys of the Monongahela and Allegheny Rivers. Elsewhere, the preexisting natural environment was converted to funnel investment in the form of raw materials, machinery, and factory structures into industrial complexes.

As many of the industrial suburbs examined in this volume show, industrial activity coalesced around preexisting commercial, market, and industrial satellite towns within twenty-five to forty miles from the city's central business district. Pullman, one of the major industrial areas of Chicago's Calumet district after 1880, formed around the antebellum development of Roseland. Baltimore's metropolitan fringe was littered with small, water-based mill towns such as Woodberry that became magnets for large-scale industrial development. The small Toronto manufacturing suburbs of Weston formed after World War I around a nineteenth-century rural service town.[25]

Indeed, many nineteenth- and twentieth-century industrial suburbs had longer histories than many areas within cities. The origins of Lachine, a working-class industrial suburb of Montreal, for example, can be traced back to the seventeenth century. It began as an entrepôt for the fur trade, connecting the northern reaches of present-day Canada to London and Paris, the imperial metropolises of the day. With the fur trade's declining importance from the late eighteenth century, Lachine's economic activity languished for nearly a century. Revival came in the 1880s with the influx of American and Canadian corporations that constructed large steel, metalworking, and glass factories. A focal point of metropolitan Montreal's western metalworking complex, Lachine remained an independent suburban entity until very recently and reached its zenith between the world wars, long before many areas of present-day Montreal were constructed.[26]

Similarly, the industrial city of Paterson, a New York satellite city, dates from the 1790s, when Alexander Hamilton's Society for Establishing Useful Manufactures decided to lay out a manufacturing town. Although the project collapsed, the New Jersey town grew, and by the early nineteenth century it had acquired an important industrial base centered on textiles and supplemented by machinery and locomotives. As the tentacles of metropolitan New York City reached farther out, Paterson, with its dependence on downtown New York's financial institutions, wholesale houses, and large markets, became absorbed into its ambit. Like Lachine and the southern suburbs of Chicago, it was one of the many older suburban areas built up long before many of the newer zones within the city were developed.[27] In other words, identifying the new with the suburb and the old with the city fails to capture the complexity of urban form and development.

As suggested by a small body of research over the past 100 years, and borne out by the papers collected in this volume, industrial suburbs were part and parcel of metropolitan development from as early as the mid-nineteenth century. Industrial suburbanization and the formation of industrial suburbs were not unusual or exceptional developments before World War II. While the authors of this volume extend this earlier work, several important areas still need to be examined. For one thing, we need a framework within which to understand industrial behavior, industrial suburbanization, and industrial suburbs in the broader context of metropolitan city building. This would involve taking stock of the qualitative studies of real and tangible industrial suburbs made by scholars over the past 100 years.

Such a framework needs to be embedded within a rigorous conceptual account. As suggested earlier in this chapter and also in the following one, even though the "political economy approach" of the 1970s took some false turns, and was weakened in particular by the absence of research on real places and its polarization of economic activity and metropolitan space, it continues to provide a powerful framework for explaining industrial suburbanization. While the following chapter's main aim is to show how industrial suburbanization has been theorized since the early 1900s, it does point to three elements—geographical industrialization, property development, and political control—critical to a reinterpretation of metropolitan/suburban theory.

Building on this last point, greater consideration needs to be given to the flow and fixity of industry, finance, property, and social capital within the metropolitan area. Industrial suburbs are constructions. They are places formed out of the decisions that manufacturers, bankers, developers, politicians, and others make about where to invest capital. We know little about who made these decisions or about how and why they were made. The case studies in this volume provide some important leads, but much remains to be discovered about the extent and character of metropolitan manufacturing networks of capital, labor, information, and products. The metropolis was not divided into two realms—the city and the suburb—that operated separately. It was a functional economic entity held together by an extensive set of networks that crossed the boundaries of city and suburb. An examination of the processes driving industrial suburbanization and producing the suburban industrial base would help delineate the character of the metropolitan economy, untangle the differential impact of changing flows of capital investment and industrial restructuring on the urban periphery, and answer questions about the development of the multinucleated metropolitan district.

This is true also for the industrial structure of the metropolitan fringe. The studies in this volume of large metropolitan areas in California, the Manufacturing Belt, and central Canada provide insights into the suburban industrial base. A more systematic description of the types, scales, and production strategies of firms locating on the fringe, however, would furnish the basis for a more comparative understanding of the suburban economy. The studies here suggest that there are large differences between metropolitan districts in terms of the rate and character of industrial suburbanization. In Toronto, for example, the degree of manufacturing activity in independent suburbs appears to be much lower than in the other urban areas. We have some knowledge of why this may have been the case, but we need to know more.

Similarly, suburban economies vary quite dramatically. There can be little doubt that suburban manufacturing districts consisted not only of large, autonomous firms. They were also composed of firms from a variety of backgrounds that functioned within a complicated set of local, metropolitan, and regional networks. The timing of suburbanization also differed. Once again, as these essays show, we know that men, women, and children labored in large numbers in a variety of suburban workplaces long before the well-documented large-scale decentralization of manufacturing and retail jobs after World War II. But we need to answer other questions: what was the actual pull that suburban firms had for workers? And vice versa? What was the relationship between firms settling on the periphery?

Finally, the relationship between the industrial and the social character of the industrial suburbanization process needs to be better understood. Greater focus needs to be placed on the relationship of industrial suburbanization and the journey to work, the ethnic structure of particular areas, the urban labor market, and the character of the housing market. This is particularly related to the role of the property industry in the formation of suburban manufacturing space and how this was linked to transportation corridors, land speculation, and local political alliances. More also needs to be known about the boundaries operating within metropolitan areas. Was the line between city and suburb a valid one? If so, to what extent and in what ways? Research suggests that the boundary's importance for understanding the metropolitan geography of manufacturing employment and how urban space has been used by economic functions has been overstated. But more research is necessary if we are going to have a strong grasp of the shifting character of urban boundaries and the effects this had on the social and economic life of the metropolis.

2

Beyond the Crabgrass Frontier

Industry and the Spread of
North American Cities, 1850–1950

RICHARD WALKER and
ROBERT LEWIS

Cities do not grow by accretion or by the obtrusion of excrescences at the periphery, but by the establishment of nuclei in the penumbra and the gradual filling in of the interstices between the nuclei.

—E. M. Fisher

The conventional story of suburbanization in Canada and the United States portrays an outward movement of residences from the cities that only lately has been fueled by the dispersal of employment to the urban fringe. In the classic studies, suburbia is conjured up as an image of "homes in a park," a middle landscape constituted as a way of life halfway between city and country.[1] This conventional wisdom needs considerable revision. Residential areas have not singularly led the way outward from a previously concentrated city, but have always been joined at the hip by industry locating at the urban fringe. The outward spread of factories and manufacturing districts has

Reprinted with minor revisions from Richard Walker and Robert Lewis, "Beyond the Crabgrass Frontier: Industry and the Spread of North American Cities, 1850–1950," *Journal of Historical Geography* 27 (2001): 3–19, by permission of the publisher Academic Press.

been a decisive feature of North American urbanization since the middle of the nineteenth century. Suburban growth as a whole has been a mixture of industry and homes, the city sprawling ever outward from its initial point of establishment and repeatedly spilling over political, social, and perceived boundaries.[2] The result has been extensive multinodal metropolitan regions. In this essay we present a theoretical reinterpretation of industrial suburbanization. We argue that industrial decentralization has been repeatedly misinterpreted as new and unprecedented, rather than recognized as an extension of past trends. In contrast to the prevailing interpretation, we claim that industrial suburbanization is the product of a combination of the economic logic of geographical industrialization, investment in real estate, and political guidance by business and government leaders.

Industrial Concentration and Residential Suburbanization: The History of an Idea

It took the Progressive Era reform and Settlement House movements to spur serious study of the internal structure and expanding scale of American and Canadian cities. Even though urban reformers focused on housing and social conditions, and ardently believed in the benefits of "decongestion" from the inner city to the rapidly growing suburbs, they documented industrial dispersal and pointed to the emergence of metropolitan urban form.[3] But this was forgotten after World War I, when urban research became academically established through the work of urban sociologists at the University of Chicago.[4] The Chicago School's "ecological" model focused on the urban core, the distribution of land uses around the center, and the sequence of land-use change as the city expanded. Unfortunately, this model set the priority of social geography over industrial location in urban studies, fixed the image of land-use rings, emphasized segmentation rather than unity of employment and residence, and established the idea of city growth as a process of decanting the core. The leading study of suburbia in the 1920s similarly enshrined the notion of residential periphery and industrial core.[5]

The massive study of New York for the Regional Plan Association, which outlined the movement of industry to the fringe and the development of a multinodal metropolis, was much richer in many respects, but it lacked an interpretative theory.[6] The Depression era saw another round of studies on the size and expansion of cities, industrial zones, and industrial dispersal. Extensive research charted the decentralization of industry to the suburbs, modified the ecological model to allow for "wedges" and "nodes" of land uses expanding outward, and laid out the parameters of the urban property market and its cycles.[7] But the Chicago School notion of business core and residential rings was not easily dislodged.

After World War II, a new wave of urban studies appeared. Extensive residential suburbanization again grabbed the spotlight after 1945, despite impressive industrial dispersal during the war.[8] A spate of books appeared, treating the postwar suburban push as unprecedented in the same way that Progressives had hailed the suburban trend in their day.[9] The premise had changed in one remarkable way, however: the central city was now seen as endangered by the pace of suburbanization.[10] The Regional Plan Association commissioned another massive study of the New York region to assess the viability of Manhattan; this had a greater impact on urban research than its predecessor because it retooled the theory of agglomeration to fit office activities at the center, and the theory of "industrial maturation" to explain dispersal.[11] All the same, the Chicago School model continued to dominate discussions of cities, and economists and geographers who worked on urban location in the banner years of the 1960s busily constructed formal models that posited centrality in the manner of Park and Burgess or the recently rediscovered von Thunen.[12]

In a carryover of intellectual momentum reminiscent of post–Progressive era scholarship, the 1970s and early 1980s witnessed a flood of studies of urbanization and suburbanization. While some scholars recognized the role of industry in urban decentralization,[13] most began with the same stylized facts about the central location of manufacturing and rings of residential land drawn out to the suburbs.[14] By the end of the 1980s, a new generation had proclaimed the emergence of a shocking phenomenon called, variously, Exopolis, Postsuburbia, or Edge City. These new employment centers at the metropolitan rim—the product of a decade of booming growth, property speculation, and large-scale development, with concomitant dispersal of industry, offices, and retail malls—were treated as something entirely new rather than as the latest episode in a long-running story of North American urbanization.[15]

The Conventional Logic of Industrial Dispersion

Once the outward flow of industrial sites and employment is allowed into the center of the picture, the explanation for metropolitan expansion must be drastically altered. In order to rethink our theory, it is necessary to revisit the reigning model of centralization and decentralization of industry and employment. The conventional view begins with the assumption of overwhelming concentration of industry in the urban core in the nine-teenth and early twentieth centuries. While the rate of decentralization has been debated,[16] virtually all students of urban employment would agree with Allen Scott that centralized production was "characteristic of the large metropolis well into the twentieth century."[17] According to most urban scholars, industrial suburbanization of any significance did not occur until after World War II.

The principal factors behind industrial centralization in traditional intra-metropolitan location models are transportation costs and agglom-eration economies. The movement of industry to the periphery, in this view, only came about with recent advances in transportation systems, industrial process technologies, and business organization that lowered the cost of locating away from urban centers, reduced the effects of agglomeration, and liberated factory and firm from the urban land nexus. The traditional emphasis on transport costs and agglomeration effects in urban land-use models follows the theoretical lead of Alfred Weber.[18] For most writers, the central manufacturing zones result from the minimiza-tion of transport costs to the urban market and to centrally placed ship-ping nodes such as ports and railway depots.[19] This skeletal explanation is fleshed out with a theory of economies of proximity among many small firms concentrated in a limited area. Different versions of this account exist, but for Weber it is primarily transportation cost reduction among all firms that explains clustering and, secondarily, access to a centrally located labor pool. The converse of this theory of concentration is the transport-driven model of industrial decentralization. In the classic ver-sion, cars and trucks lower costs of transport dramatically over rail and water and lessen dependence of urban manufacturers on ready access to central rail and harbor facilities. For example, "between 1915 and 1930, when the number of American trucks jumped from 158,000 to 3.5 million ... industrial deconcentration began to alter the basic spatial pattern of metropolitan areas."[20] Transportation becomes virtually universal in explaining suburbanization based on the argument that cars and trucks provide unprecedented speed and flexibility in moving workers and goods.

The product-cycle model was grafted onto Weberian location theory, adding the idea of industrial "incubation." The central city, from this per-spective, has a relative advantage as a source of innovation, thanks to max-imum access to markets, new ideas, skilled labor, and finance. It serves to incubate new products and new firms, which subsequently move to the suburbs (or backward regions).[21] Agglomeration loses its grip as industry matures; the shift to mass production eliminates the reliance of small, specialized producers on each other by standardizing input-output linkages and bringing a range of activities into large, integrated factories.[22] The result is the dispersal of firms from the core to the city fringe.

The last addendum to the conventional theory of industrial decentralization allowed for the evolution of business organization from small, single-plant firms to the modern corporation. Theories of "corporate location" absorbed the product cycle into an overarching theory of the dispersal of branch plants to peripheral regions and countries from a corporate core.[23] The causal mechanisms are standardized, large-volume flows of inputs and outputs, large-scale plants, and internalized transactions. The corporate umbrella severs the enterprise from external linkages (commodity trade, specialized labor skills, management inputs, etc.), breaking the collective logic of agglomeration and freeing corporate-owned factories to seek cheap land and labor far from the city.

Urban Geographical Industrialization

The conventional explanation of industrial location in city and suburb has serious problems. In the first place, transportation limits but does not determine the location of industry. Undeniably, transport costs influence the geography of urbanism: industry has always clustered near transportation nodes and corridors, whether harbors, rail lines, or highways, in a way that leads to transport-tied corridors of industrial land use.[24] Over time, improvements in transport have also allowed the city to spread out, but transportation access has been more widespread than conventional models allow, for three reasons. First, nineteenth-century transport modes were not as fixed or nodal as is commonly asserted; extensive water and rail systems surrounded cities and even penetrated the countryside. Second, water and rail systems could be brought to industry as well as industry brought to them, through investment and spatial extension; transport access is often the *dependent* variable in the equation of industrial location. This applies to a specific rail siding or an additional dock, as well as to the construction of entire canals or rail lines. Third, trucks did not suddenly revolutionize location, because industrial dispersal using wagons, railroads, and boats took place before and after the truck.[25]

Weberian theory, including its account of agglomeration, suffers from undue emphasis on cost minimization with respect to input factors. Demand conditions are important, but even though capitalist firms try to keep costs down and weigh the relative prices of inputs, industrialization is not principally an optimization problem; it is a dynamic process in which new commodities and new ways of doing things continually displace the old, and today's prices, based on further technical change, displace yesterday's costs.[26] The drive to improve productivity through standardization, rationalization of work flow, mechanization, and automation is essential to the history of modern industry, beginning with the Industrial Revolution.[27] In such a context of rising productivity and increasing returns to all factors, optimization models simply do not work.[28]

Product cycle–incubator theory adds a needed element of innovation and productivity advance, but it does so in a highly stylized manner that assumes, wrongly, that the new is small and the old is large, that industries mature in a systematic way, and that they are well behaved in their locational choices. The evidence thus is mixed and quite often contradicts the model. Jewelry firms, for example, are small and centrally located but neither cost-sensitive nor innovative, and they do not grow up to be like auto plants. Refineries and shipyards were always relatively large. And while large factories were often pioneer dispersers, they can also frequently be found functioning quite well very close to the center of big agglomerations. In either case, they were prominent in cities well before the twentieth century. Conversely, those industrial activities most prone to clustering—small workplaces and craft-like production—are by no means always to be found in the central city. Finally, outlying locations are frequently the first implantation of a new activity in the city, and the incubation stage is skipped entirely, time and again.[29]

In short, industrial development and location is not a monotonic, uniform process. What we see instead are successive eruptions of new industries, embodying new products and new technical bases, and a diverse array of production formats evolving and restructuring over time. Technical change has developed on a variety of material bases in different industries, has moved along divergent industrial (and company) trajectories, and has been altered radically by new discoveries from time to time.[30] This has meant many patterns of initial location, agglomeration, and dispersal, giving North American cities quite distinctive industrial foundations, patterns of uneven spatial expansion of those sectors, and episodic additions of wholly new industries to the mix.[31]

In fact, industry and the city have grown together as a unified process of geographic development. Industry does not locate in the city; it helps create the city.[32] Urban expansion is based on the ability of industrialization and capital accumulation to create places at the same time that they make commodities, build factories, raise up a labor force, and introduce new technologies. This process of "geographical industrialization" has the following principles.[33] First, new industrial locales are able to break away from old centers and existing economies of agglomeration, thanks to both the rapid rates of accumulation and the experimental nature of their growth process. They are likely to avoid existing concentrations if they fear the effects of established labor practices, management outlook, or worker militancy. Second, growing industries build up extensive territorial concentrations of related activities, such as specialized suppliers, merchants, financiers, and educational institutions, and spin off new firms and even new industries in their process of expansion. Third, new industrial implantations attract and train new labor forces, steeped in the particular ways of working, technology, and ethos of the industry. These fresh labor forces may have little in common with other segments of the labor market. Last, given the repeated and permanent nature of industrial revolution under capitalism, the space-economy has undergone many changes and upheavals.[34]

Applied to the urban arena, these principles suggest that as cities develop new industrial sectors or as existing industries restructure and expand, successive nodes of growth erupt in outlying areas, growing in time to fill up the neighboring suburban territory. As cities have grown, layer upon layer of suburban development has been added to the built-up area, leaving former outlying districts well inside the metropolis and often erasing in the process historic patterns of expansion by dispersion. After many years, it is easy to mistake the older edge cities and secondary nodes for part of a single "central city." Modern metropolitan areas are so huge that even large and distant suburban edges of the past, such as Brooklyn, Oakland, or South Chicago, are now deeply embedded in the structure of the city. The study of North American urbanization thus requires a model that begins with the simultaneous march of industry and cities outward, rather than a two-stage process of building a dense concentration of activities in the core in the nineteenth century and then decanting them in the twentieth.

Industrial Districts and the Multinodal Metropolis

The process of urban industrial growth has another crucial dimension besides the outward flow and build-up of the city: the appearance of distinctive industrial districts within a multinodal metropolitan area. Classic agglomeration theory does not explain this phenomenon; the city is a single generic agglomeration with industry confined to the core. Conversely, traditional decentralization models allow only for the dispersal of large factories under the umbrella of the modern corporation. In both cases, too much is missing from the real fabric of urban industrialization; to recover it, we must consider the problem of industrial organization and the spatial division of labor.

Industrial organization has come under renewed scrutiny in recent years, and the older view of universal evolution from small firms to large corporations is no longer viable. Organizational forms are many and varied, across sectors, places, and time. Small firms recur and persist while large companies show considerable variety in internal makeup. Big corporations do not simply insulate themselves from the market but interact with it, and with other firms, in a more or less open manner, depending on strategy and circumstance. Market relations can be attenuated in several ways as well. Finally, firm and market are not the only forms in which the social division of labor is integrated; territorial aggregation and local governance systems, local and national states, industrial associations, and other organizational tools also play a part.[35]

The most sophisticated model of urban industrial clustering is that of Allen Scott, who tries to capture the dynamics of industrial agglomeration and decentralization in terms of "transaction costs." Oliver Williamson developed the theory of transaction costs to translate Alfred Chandler's insights into the rise of the modern corporation, based chiefly on the technical imperatives of scale, into the language of neoclassical economics. Scott realized that the same insight could be applied to geography, allowing for a reworking of Weberian agglomeration theory. He argued that urban concentration provided an alternative to the large firm. Complexes of vertically disintegrated producers within specific industrial sectors cluster to take advantage of mutual interaction. Complexes grow through the intensification of the division of labor, multiple linkages among firms, and flexibility in the face of changing markets; these generate economies of scope for individual specialists and collective economies for the entire industry.[36]

Scott's work complemented that of European researchers who examined the vigorous industries of the Third Italy and rediscovered Alfred Marshall's idea of the "industrial district."[37] While initially arguing for a small-firm model of clustering, Scott realized that both large and small factories and companies are embedded in industrial districts.[38] Size would be decisive if external exchange were the only reason for agglomeration, as early transactions models implied. Yet the benefits of interaction go to the heart of all extensive divisions of labor because they lower costs of interaction among dependent parts of production systems, reduce the risk of investment, lessen turnover time, and offer institutions of collective governance. Furthermore, they offer dynamic advantages by stimulating the collective process of learning and providing a milieu of problem solving and innovation.[39] Scott thus abandoned the simple model of central agglomeration and decentralization of large factories to the suburbs in favor of one of multiple clusters throughout the metropolitan region; industrial districts can occur in any number of high-tech, large-batch, or "new craft" sectors.

It can be seen, on further reflection, that industrial districts were an essential element of American cities from early in the history of industrialization. By the mid-nineteenth century, a system of dense industrial districts was embedded throughout the Philadelphia metropolitan area, for example. Boston contained a set of distinct industrial suburbs specializing in such products as shoes, machinery, and textiles, and a distinct set of manufacturing districts quickly developed in such cities as Baltimore, Montreal, Toronto, and Los Angeles.[40] If these districts had once been close enough to the center to be confused with a single manufacturing core, by the turn of the century urbanization had reached the metropolitan scale. Since at least 1850, the North American city has grown largely through the accretion of new industrial districts at the urban fringe, becoming multinodal in the process.

The basic theory of industrial districts, however, begs the question of geographic scale; and this is crucial to the understanding of metropolitan areas. At what scale are the forces of agglomeration operative, and how far do their relations of mutual dependence and benefit extend into the world beyond the plant gate? These difficult questions have only begun to be seriously examined in discussions of the dialectics of the local and the global.[41] Large cities and metropolitan areas are units of effective interaction and agglomeration in their own right, as well as assemblages of industrial districts. Furthermore, spatial concentration and dense geographic networks of interaction also can be observed at enormous national and continental scales, running far back in the record of European and American urbanization.[42]

The metropolis provides a connective tissue embracing both individual plants and sets of industrial clusters. The specialized "industrial district" may effectively be the whole metropolis, such as the immense center for steel production that emerged in Pittsburgh, but it is inevitably "lumpy," owing to the presence of industrial districts and subregions. The benefits of urban proximity cut across industrial sectors; they do not necessarily depend, as Scott's theory proposes, on intimate relations of vertically disintegrated plants operating within a single sector. The interplay of economic activities can be fruitful across extensive divisions of labor.[43] Even as specialized an industrial city as Pittsburgh (steel) or Detroit (autos) is a creature of many parts, and most submetropolitan industrial districts also embrace more than one industry and commonly shade into larger penumbras of localization within the metropolitan area. Boundaries are slippery, and interactions highly diverse. As the city grows, so does its spatial division of manufacturing; each new industrial zone and its mix of industries forges a niche in the expanding metropolis, at the same time that it adds a new dimension to the fabric of the metropolitan built environment.

The geography of labor markets makes up a key component of the multinodal metropolis, as Weber and subsequent agglomeration theorists have recognized. The city is one large labor market, but, at a finer scale, metropolitan areas embrace many nodes of industry and fields of workers' residences, linked by transit and daily journeys to work. Industries come with distinctive labor demands and labor relations, and these are spatialized in local labor submarkets.[44] The internal geography of cities and metropolitan regions thus is marked by the impress of the division of labor and labor market segmentation, as well as by larger class and racial divides.[45] This sorting results from many decisions by employers and workers about where to operate and to live, and by their jockeying for advantage. For example, in order to reduce turnover and labor militancy and to increase job-specific learning and identification with the company, managers often attempt to create a specialized, isolated labor force close to the plant or district. A suburban locale gives the employer arm's-length access to the urban labor pool, thereby avoiding the "corruptions" of working-class life and mobilization. A well-oiled employment relation with a subset of workers in an isolated location, frequently on the fringe of the metropolis, may outweigh the benefits of a large urban labor market. The capitalist makes a double calculation of labor markets, at the plant or district level and at the metropolitan level. This labor-relation calculus both fragments and disperses the metropolis, as employers try to carve out space for their own protection and exploitive aims within the larger urban field of mutual interaction, labor assemblage, and cross-commuting. This logic is in evidence in places as diverse as Homestead and Pullman in the 1880s and Lakewood and Fremont, California, in the 1940s and 1950s.[46]

Building Out the Metropolis

Recent literature in industrial geography advances our understanding of spatial concentration and dispersal significantly beyond the old models of urban centrality and suburbanization, but it has not made the further link to the build-out of the city. The way urban areas expand through the mediation of property developers pursuing their sectoral logic of investment, production and profit must be examined, because a principal dimension of urban industry or "production" is the construction of the city itself. Property capital's imprint on the suburban landscape can be discerned in various ways, including the shape of lot sizes, building placement, construction type, infrastructure, and improvements. This contributes to the array of urban forms that constitute the everyday vernacular landscape of the city, as well as to striking elements of homogenization across the North American urban system.[47] The urban mosaic, or the mesogeography of urbanization, has four critical elements: property developers, building cycles, financial speculation, and uneven development. They combine to produce the repeated eruptions of new constellations of employment, transport, and residence at the metropolitan fringe, and the great swaths of construction laid down in the form of peripheral belts, jutting wedges, industrial districts, satellite towns, and edge cities.

Cities have always grown at their edges, but it is erroneous to think that suburban industrial spaces, any more than residential areas, are built on demand without regard for the profits to be made from investment in land. The commodification of land, property investment, and speculative building have been hallmarks of urbanization and national expansion in the United States and Canada.[48] Property investment at the suburban fringe creates the possibility of enormous gains through the maximization of the returns of capitalized rent. Because profits increase with distance from the fringe, the search for profits in land speculation by property investors, developers, and financial institutions exaggerates the demand for peripheral sites, tempting industry and residents to move to the suburbs and pulling the city outward into the space-extensive form characteristic of North American cities.[49] This holds even for the industrial company acquiring land: suburban sites have offered not only low prices and easier assemblage of large plots for factories, but also the promise of speculative profit if surplus acreage is sold or developed.[50] The property industry, moreover, has been particularly inventive in creating complete urban environments, from the housing tract to the regional mall to the industrial park. These condensed pieces of urbanity can be set down in the greenfields like seedlings, helping the city take root more quickly in fringe areas.[51] To make sure investments are realized, promoters try to leverage urban infrastructure and other investors outward in order to "ripen" their investments. In this way, the extension of industrial space has been propelled outward from the city center.

Waves of investment in property development that correspond to waves of capital investment, job creation, and surging economic activity are another essential force in metropolitan expansion. Urban growth is neither incremental nor continuous in space and time but occurs in bursts. The urban land market is notorious for boom-and-bust dynamics in subdivision, financing, and construction, with well-documented twelve- to twenty-five-year swings in activities such as aggregate building and transport expansion. This space-time rhythm appears as rings of building activity laid down around cities with each investment boom.[52] The proximate mechanism generating such property cycles is the adjustment of supply to demand that overshoots because buyers and sellers of real estate compete fiercely and time lags exist between initiation and completion of building projects.[53] Technological and design changes in buildings, infrastructure, and large-scale developments further modulate and accelerate the industrial land process at the urban fringe. The push of capital into real estate investment due to the buildup of surplus and fictitious capital in the financial system, however, is the most dramatic aspect of property booms; this has been exaggerated in eras of financial frenzy in the economy at large such as the 1920s or the 1980s.[54] Key actors are likely to reside in the core city, but the large banks and financiers have always been complemented by upstarts in the suburban fringe who fatten on property development and by outside investors and lenders from other big cities, even other countries. While little historical documentation exists about investment in, and financing of, industrial suburban sites, the speculative processes in industrial and suburban land probably are similar.

Finally, city building through industrial suburban growth occurs within an economy that demonstrates persistent unevenness in rates of growth and capital accumulation among different industrial sectors and places. Capital flows triggered by unequal and fluctuating rates of profit and accumulation in the larger economic system fuel industrial shifts and property booms.[55] As a result, in places where investment surges into new industrial suburban development, great swaths of cities can be laid down in short order before the hand of capital moves on. These temporal-spatial dynamics of capitalist growth have shown up clearly in the metropolitan record since at least the late eighteenth century in North America.[56]

Politics and Planning of Industrial Suburbanization

In addition to the economic logic of industrialization and property development, political intervention and conscious planning have also played a significant part in the intentional process of shaping and reshaping North American cities. Despite the apparent chaos of urban building, a prevailing vision of urban expansion and suburbanization has guided the plans of industrialists, developers, and governments. The construction of cities is more than an exercise in economics; it is irreducibly about the search for geographic control, or the politics of space. Industrialists and other capitalists are acutely aware of the contradiction between the concentration of people and industry in the city and keen to maintain their prerogatives in the arenas of investment, work, and profitability, now termed the "local business climate." Location at the suburban fringe and outlying districts has offered the hope of combining the manifest benefits of access to the city and its agglomeration economies with a degree of freedom from the working class, city politics, and contending business interests. Because agglomeration effects can be created in outlying districts within reach of the urban center and are operative at the metropolitan scale, this political elbow room can be created by means of industrial suburbanization and a space-extensive, multinodal city form. This pattern was worked out during the nineteenth century and put decisively into place after the turn of the twentieth.

Industrialists began moving large-scale operations to the urban edge in the 1800s. These sites were often beyond the city limits, but they tended to be absorbed as municipal boundaries expanded. Jurisdictional inclusion in the city was usually prerequisite for much-needed infrastructure for both factories and worker housing. In the nineteenth century, cities provided government at a level of service and competence higher than other units, and business found ways to influence decisions despite the popular mobilizations of ward politics. Indeed, local bourgeois alliances were extremely successful at garnering local, provincial, state, or federal funding to build key infrastructure and utilities such as canals, sewers, and harbors.[57] These developments often underwrote suburban nodes and enabled industrialists to consider moving their plants to the metropolitan outskirts.

By the end of the nineteenth century the level of labor militancy and political upheaval associated with reform movements had risen dramatically. Capitalists became increasingly uneasy about their control of urban geography, and the politics of urban space became a subject of intense debate.

Discourse both on the evils of "urban congestion," labor militancy, political corruption, and moral turpitude, and on the virtues of the suburban solution to the dense city form, became so heated that the viability of the labyrinthian spaces of big cities was thrown into question. The result was the rethinking by the capitalist class of its economic behavior and the growing desire for escape to the suburbs among the better off. Industrial dispersal could be seen, thereafter, as not only good for business but as a social virtue and even a necessity to ward off revolution and degeneracy in the body politic.[58] The attractions of decentralization increased correspondingly, and new outlying industrial sites began to multiply.

Planning was the handmaiden of politics in helping to create and shape suburban industrial space. The most limited form of planning for industrial sites at the urban periphery is the private assemblage of land for that purpose. The company town, embodied in Lowell, Pullman, and Homestead, was an early form of planning undertaken by a single company with a vision of housing provision and proper social life for "the hands," but it was expensive and usually found to be less conducive to labor peace than the distractions of urban life.[59] The industrial park is another basic form of planning; land is carefully prepared, provisioned, and preplanned by the developer, in concert with local authorities.[60] At an even larger scale, entire industrial suburbs, such as the Chicago stockyards or South San Francisco, were carefully planned as joint development efforts between industrialists and suburban governments.

As suburban jurisdictions proliferated after the turn of the century, many aimed to attract industry, most worked hand in glove with real estate promoters, and virtually all tried to provide the best business environment money could buy. Dozens of suburban governments around every big city became suppliers and boosters of industrial land away from the urban core, often marketing themselves shamelessly and offering subsidies to capture new investors. In some cases industry could wrap itself in the cloak of specialized city government, such as West Allis, Gary, Vernon, Emeryville, and Maisonneuve, and turn its back on the exactions of civic politics and social demands for revenues and responsibility.[61]

Suburban governments took up the call for administrative reform with a vengeance. Most suburbs reduced the power of elected officials and political appointees by installing professional city managers, planning boards, and public works departments to provide for industry and urban infrastructure—all of which were nominally independent but worked closely with business.[62] In mixed-use suburbs, zoning arose at the turn of the century as a form of spatial ordering by means of local government.

Supplementing building regulations and covenants, zoning became a mainstay of land planning and set-asides for industrial expansion by the 1920s. While zoning was used to protect residential areas from encroaching industry, manufacturers could also secure their right to operate without interference by complaining and litigious middle-class neighbors. Zoning also could be conveniently altered by congenial planning boards to serve developers' interests.[63]

Industrial parks, company towns, and industrial suburbs are the clearest bounded units of planning and governance for the purpose of industrial decentralization on a scale larger than the unitary factory site. But they only partly cover the scope of planning of the suburbanizing metropolis under the watchful eye of the doyens of civic capital. The city planning movement of the Progressive Era is a well-known instance of large-scale intervention, a much ballyhooed attempt to assert formal master plans for development and public works over the seeming chaos of city growth, at both the center and the margins. Such efforts were invariably backed by a constellation of bankers, utility owners, industrialists, merchants, and land investors, led by a few visionary architect-planners and politician-businesspeople.[64] Civic leaders were forever trying to manage growth for personal and collective advantage. They readily grasped the way wedges of manufacturing and new industrial districts were pushing and leaping outward, forming satellite zones beyond the compass of the built-up city. They further understood the need to maintain the necessary links between industrial, residential, and commercial zones, and to install transportation arteries, such as canals, ports, railroads, highways, and eventually airports.[65]

Creating suitable, functioning urban spaces was an ongoing job, with key planners and boosters taking the lead; but, equally important, private investors, elected officials, city managers, and public works directors acted in concert to endorse, promote, and provision growth along the lines previously set out. Political alliances and urban growth coalitions were regularly constructed in support of the city-building effort.[66] Thus the "weave of small patterns," as Sam Warner calls the remarkable coherence of suburbanization,[67] was more articulate, better orchestrated, and more clearly envisioned by the elite than the metaphor allows, and it was effective in getting the job done within liberal political culture. One may quarrel over the degree of ruling-class coherence and manipulation at work—Cochrane and Miller declare that "it is impossible to exaggerate the role of business in developing great cities in America"—but, without question, considerable effort went into steering the whole process from on high.[68]

At the burgeoning edges of the metropolis we find a full panoply of workplaces, homes, infrastructure, and commerce that make up the economy and life of the city. These suburban nodes have ranged widely in size, character, and relative autonomy from the parent city, depending on circumstances of economic base, class base, political history, and the like. These extrusions of the growing city are not altogether random, but the complexity of metropolitan expansion requires the kind of nondeterminate, nonuniformitarian theory now associated with interplanetary geophysics or hydrodynamics.[69] There are no "typical" cities and suburbs, no uniform growth paths, no easy way out of the particularities of history; nonetheless, it is possible to capture the major forces at work behind diverse outcomes. We argue that the combination of geographical industrialization, land development, and metropolitan politics and planning is a theoretical framework that offers a means to advance beyond previous theories at the disposal of urban scholars.

3

The Emergence of Industrial Districts in Mid-Nineteenth-Century Baltimore

EDWARD K. MULLER and PAUL A. GROVES

arge North American cities displayed a geography in transition during the middle decades of the nineteenth century. Although commerce continued to dominate economic life and intra-city movement remained largely pedestrian and horse-powered, rapid population growth, the influx of large numbers of foreign-born immigrants, and the initial phases of commuting by omnibus, horsecar, and steam railroad began to alter the traditionally compact, dense, and heterogeneous mixture of land uses of the geography of the early American city.[1] Especially important for this changing geography was the growth of new types and organizational forms of manufacturing.[2] Traditional artisan shops, small manufactories, trade-oriented raw material proces-

Reprinted with a revised introduction and minor editorial changes from Edward K. Muller and Paul A. Groves, "The Emergence of Industrial Districts in Mid-Nineteenth Century Baltimore," *The Geographical Review* 62, no. 2 (1979): 159–78. Courtesy of The American Geographical Society. Historical geographical scholarship since 1979 suggests that "commercial," rather than "mercantile," is a more appropriate characterization of the city in the first half of the nineteenth century. Because of Paul A. Groves's untimely death, all changes are the sole responsibility of Edward K. Muller.

sors, and shipbuilders persisted, often in large numbers, well past the Civil War. But, increasingly, some firms adopted new organizational strategies or the steam engine and relied on the railroads as customers and transporters in order to lower costs, dramatically expand production, and become much larger in value of production and number of employees.[3] The growth in the number and size of manufacturing firms led to the localization of some firms into industrial districts and the expansion of manufacturing on the urban periphery. In short, the early stages of emerging industrial districts and industrial suburbanization at mid-century foreshadowed characteristic features of the later industrial metropolis.

The geography of the mid-century transitional city is not clearly understood. Conventional descriptions of a congested, heterogeneous urban landscape frequently rest on impressionistic surveys of contemporary data and confusing, often imprecise sketches of change. Indeed, interpretations of the transformation of the commercial city into the industrial metropolis understate the patterns and processes of the transition that, by encompassing decades of change, comprise a period of urban experience capable of influencing the subsequent industrial one.[4] In characteristics such as dominance of commerce and pedestrian movement, persistence of some craft activities, and general heterogeneity of land use, the North American city was still commercial. In characteristics such as population size and composition, embryonic phases of commuting, and the emergence of large-scale and producer-goods manufacturing, the mid-century city displayed the industrial form. The changes in some sectors of industry, involving increased production, large numbers of employees, and often sizeable parcels of land, contributed to this evolution of urban geographical structure.

In this essay, the location of all industrial activities in 1860 for the city of Baltimore, a large commercial city experiencing the various mid-century changes, is examined in detail and subsequently compared with the 1833 commercial pattern. We argue that the new and large-scale manufacturing activities formed the basis of emerging, separate industrial districts that were distinguished from one another by type of product, organization of production, power source, or composition of labor force. Some of these districts appeared at the edges of urban development, in effect leading the outward thrust of the city. Although this growth of industrial districts only modestly expanded the city's spatial periphery, it did establish areas of industry, which would influence the location of future waves of industrial suburbanization. Moreover, these peripheral industrial districts tended to rely more on steam power and attract more factory-oriented, railroad-dependent firms than the districts within the older commercial city did. This locational specialization at mid-century resulted in districts of distinctly different industrial employment opportunities.

Urban Manufacturing: Organization and Pattern

Although there have been few specific studies of the location of manufacturing activities, the broad outlines of their distribution in mid-century cities are generally agreed on in the literature. New kinds of industrial activity and consequent distributional changes existed simultaneously with the older forms and patterns of the antecedent commercial city. In the latter, manufacturing activities supported commercial functions and provided goods for the local inhabitants and regional hinterland population. Processing of trade commodities was a major category of manufacturing that included grain milling, distilling, tanning, meatpacking, and sugar refining. Shipping frequently generated shipbuilding, repairing, and outfitting activities. Moreover, the growing city population required industries to produce building materials and consumer goods. Artisan workshops were pervasive, and accordingly most manufacturing establishments were small, unspecialized, and dependent on hand labor.[5]

The location of manufacturing in the commercial city reflected the pedestrian basis of the circulation of people, goods, and information. The commerce-serving function of many industries and the inefficiency of handling bulk goods produced an agglomeration of warehousing and manufacturing along the waterfront.[6] The need for proximity to information sources and inter-industry linkages enhanced the attractiveness of the central wharves and resulted in some clustering of establishments.[7] Rising land values, however, forced some industries using large parcels of land, such as shipbuilding and foundries, to peripheral areas of the waterfront.[8] At the same time, many artisan workshops preferred the central area for its accessibility to the entire urban market. With the rapid growth of the city, consumer-oriented workshops tended to separate from the wharves and to move into an emerging business area. Thus the waterfront and central core were characterized by a marked concentration and a mixture of small-scale manufactures, although pressures for sorting within these areas had begun in the larger commercial centers.[9]

Other areas of the commercial city also contained manufacturing. The coincidence of workshop and residence combined with the demand of the local population for convenience goods to disperse artisan shops throughout the city. Industries that required large land areas (rope walks) or specific raw materials (brickyards and quarries) and industries with noxious qualities (slaughtering and tanning) were located on the urban periphery, frequently part of fringe zones of land use.[10] Because flour mills and sawmills depended on waterpower, they commonly created satellite communities around waterpower sites well beyond the urban area. The development of water-powered textile mills, particularly during the first third of the nineteenth century, fostered the growth of separate satellite industrial towns throughout the Northeast. Despite this variety of locations, the large majority of industries were concentrated in the central area, while the remaining establishments were generally small in scale and scattered about the urban area. Within the compact commercial city, substantial industrial districts had not yet formed.

The demands on manufacturing after 1830 remained essentially those of an expanding agricultural and commercial society that required improved productivity in farming, in processing, and in transportation.[11] The persistence of manufacturing concerns and traditional modes of intra-urban movement for most workers and commodities maintained a continuity with commercial city patterns. Most descriptions of the mid-century city note the heterogeneity of central-area manufacturing, the dispersion of artisan shops, the waterfront's attraction, the peripheral location of large or noxious land uses, and the neighboring mill towns.[12] Although the vast increase in the number of establishments seems to have contributed to an intermingling of different types of manufacturing, most writers also recognize that the proliferation, along with changes in the composition of labor, type of product, power source, and organization and technology of production, effected some departures in the location of urban manufactures.

The elaboration of canal and railroad networks after 1830 and the growth of the regional economies west of the Appalachians expanded marketing opportunities for merchants and manufacturers in the seaport cities of the northeastern United States. Those interregional demands for consumer goods necessitated increases in productivity that were met initially by organizational modifications. The traditional production unit of the commercial city—the artisan shop, which usually combined production, sales, and residence at the same location—was reorganized through the differentiation of the manufacturing process into the putting-out system and the expanded workshop (sometimes called an inside shop). Under both systems master craftsmen and merchants divided production into specialized tasks of different skill levels. That differentiation permitted increased employment of inexpensive, unskilled female and immigrant labor. In the putting-out system, skilled tasks were performed at the workshop, while the remaining operations were assigned to semiskilled and unskilled workers who performed the tasks at their residences. The expanded workshop assembled large numbers of workers, that is, between fifteen and fifty people, under one roof. These workshops were frequently located in older buildings, particularly warehouses, of the city's central area, where rents were relatively low and accessibility to merchants and to transportation for materials was excellent. Central locations also were proximate to large numbers of unskilled workers, often immigrants.[13]

Although the workshops of many industries existed throughout the city, the central orientation of many, especially in industries dependent upon nonlocal trade, added to the manufacturing character of the burgeoning central area. Sam Bass Warner indicated that a quarter of Philadelphia's industrial labor force worked in the central ward at mid-century.[14] Although he described a mixture of different types of manufacturing activities, the apparent heterogeneity may have partially resulted from his coarse spatial units of analysis. Other studies suggest that many commerce-related manufacturers still crowded along the waterfront, while local consumer-serving artisans such as tailors, cigar makers, printers, and piano makers gravitated toward the central business district of retailers, offices, public administration, and entertainment.[15] Warner also cited the growth of Philadelphia's first garment workshops in the south side slums adjacent to the central area, and David Ward noted the existence of multifunctional warehouse workshops both within and adjacent to Boston's developing business district.[16] With the increase in the number of establishments, changes in the organization of production, rising land values, and emerging central business functions, the typically heterogeneous mixture of small firms of the commercial city's central area gave way to an internal differentiation of land use.

Adoption of the steam engine and development of producer-goods industries, especially in transport-related products, also brought the factory system to the city. Although steam power was applied primarily to processing at raw material sites before 1860, the steam engine allowed factories to develop in port cities where raw materials for both energy and production could be assembled and where external economies were available.[17] Increased use of machines in some industries furthered the development of the steam-powered factory. In turn, the burgeoning demand for steam engines and machines as well as for steamboats, rails, locomotives, railroad cars, tools, and a variety of iron products generated the expansion of large metal producer-goods and fabricating industries.[18] The characteristics of large-scale production and of bulk raw materials and products forced those industries to find locations that combined ample space with minimal intra-city freight movement.[19] Because of the existing high-density development of central land, the metal industries tended to locate peripherally along the waterfront, near new railroad terminals, or at the junction of these two transport modes.[20] Railroads were late additions to most established large cities and acted as a peripheral focus for manufacturing.[21] Warner, for example, described the emergence of locomotive and metalworking establishments around Philadelphia's northwest rail yards.[22]

To summarize, the mid-century city contained a mixture of small traditional shops and larger warehouse workshops, of handicraft productions and steam-powered factories, and of core concentration and peripheral locations. The differing scales, forms of organization, and locations among the various industries created a pattern that did not conform to the well-known structure of manufacturing in either the commercial or the industrial city. Localization by new complementary manufacturing activities suggests the growth of industrial districts. Philadelphia offers evidence of the emergence of such districts, but its large population—double that of Baltimore or Boston—and advanced state of manufacturing at that time may have made it unrepresentative of other mid-century North American cities. Moreover, the absence of specificity either in the scale and proportion of such clusters of a city's total industrial activity or in the mix of industries, which would create and differentiate them, makes it difficult to evaluate their importance for the overall organization of urban activity. The study of manufacturing in Baltimore will be used to corroborate the emergence of industrial districts, as suggested by Philadelphia's experience, and to characterize more precisely their composition and significance within the city.

The Changing Industrial Geography of Baltimore

By 1860 the major northeastern seaports and Ohio Valley river cities, which had responded to new marketing opportunities, were statistically distinguishable from more traditionally oriented commercial cities such as Chicago, New Orleans, and San Francisco, by considerably greater numbers of manufacturing employees. Only much smaller specialized industrial cities, Lowell or Troy, for example, were comparable in magnitude of manufacturing employment, but these cities represented different cases in urban spatial structure because of their small populations, specialized economic base, and, frequently, company-town heritage.[23]

Baltimore's vigorous exploitation of Atlantic trading opportunities propelled the city to national prominence in the early decades of the nineteenth century. In spite of an inconsistent record after the early 1820s, international commerce remained the bulwark of the city's economy at mid-century.[24] Through regional internal improvements, including the innovative construction of the Baltimore and Ohio Railroad, Baltimore also vied aggressively for hegemony over an expanding regional hinterland as well as Ohio Valley markets. Toward mid-century, however, its merchants turned increasingly to markets in the South. Expansion of the city's trade, growth in local population, and demand from transportation companies nurtured Baltimore's manufacturing activities, most of which were originally associated with commerce. Processed foodstuffs, textiles, hats, ready-made clothing, ships, iron castings, and metalwares headed the list of items made for nonlocal markets.[25] By 1860 Baltimore had approximately 16,000 persons employed in industry, the sixth-largest industrial workforce among the nation's cities.[26] Although it had not replaced commerce and was in fact dependent on local commercial activities, industry had become an important component of the city's economy. Baltimore is therefore an excellent example of a transitional city, because by 1860 it was a leading commercial and industrial center, had substantial foreign-born immigration, and experienced rapid population growth.

The Commercial Pattern, 1833

During the first three decades of the nineteenth century, the development and distribution of manufacturing in Baltimore conformed closely to the patterns of North American commercial cities. Commerce continued to be the city's main economic activity, and by 1830 Baltimore, with a population of 80,630, was the third-largest urban center in the nation. Manufacturing was still dependent on commerce and on consumer and construction demands of the local market. No comprehensive census of manufactures exists for this period, but a gazetteer-style directory of 1833 provides a detailed, but undoubtedly incomplete, enumeration of industrial establishments from which a locational pattern may be derived.[27] This source does not contain data on either the number of employees or the value of industrial production.

Commercial city manufacturers processed commodities for foreign and domestic trades. Producers of flour, textiles, and iron goods often used waterpower and exploited sites outside the urban area. Seventy-eight mills and factories, including thirty-one flour mills, fourteen grist mills, fifteen textile factories, and ten iron and copper works, were located within fifteen miles of the city. Many of those manufacturing establishments had warehousing facilities within Baltimore's central area, underscoring their connection to the city's economy.[28]

TABLE 3.1. Number and Type of Manufacturing Establishments in Baltimore by Area, 1833

| | Industry | | | | | | Total | |
Area[a]	Ship-building, repair, and related activities	Agri-cul-tural pro-cessing	Raw mate-rial pro-cessing	Con-struc-tion mate-rials	Metal pro-cessing and prod-ucts	Artisan-domi-nated activities	No.	%
Central business area	2	2	10	0	7	125	146	39.4
Peripheral business area	1	6	9	2	6	50	74	20.0
Old Town	0	2	5	0	1	23	31	8.4
South Baltimore	1	2	4	2	0	10	19	5.0
Fells Point	2	0	0	0	0	13	15	4.0
North Falls	0	0	2	0	0	1	3	0.8
South Falls	0	5	5	0	4	17	31	8.4
Dock (A)	6	3	4	1	2	14	30	8.1
Dock (B)	1	0	0	1	4	1	7	1.9
Remainder of the city	0	1	2	3	2	7	15	4.0
Total	13	21	41	9	26	261	371	100.0
% of total	3.5	5.7	11.1	2.4	7.0	70.4		

Source: C. Varle, A Complete View of Baltimore (Baltimore, 1833).
[a] See Figure 3.1 for identification of areas.

In contrast, relatively few flour and grist mills or textile factories were located in the city, where consumer-serving, mostly artisan, activities dominated numerically. Traditional artisan-style manufacturers of consumer products, such as tailors, cobblers, bakers, tobacconists, and blacksmiths, made up 70 percent of the city's manufacturing establishments (see Table 3.1).

Simple enumeration of activities, however, obscures some changes in industrial production that were fostered by the response to interregional markets and expanding local demands. There was evidence of a few large establishments, such as Stockton and Stokes' coach-making operation, with eighty employees.[29] Moreover, large numbers of weavers apparently worked in workshops or at home.[30] Advertisements in the city directories for locally produced ready-made clothing further suggest the existence of workshop production under the direction of merchant-manufacturers.[31] At the same time, some Baltimore manufacturers had begun to use steam engines. At least thirty-two steam-powered establishments, using small-horsepower engines, existed in 1833; they included a dozen foundry and machine shops, six cotton textile and carpet factories, four planing mills, and a variety of processing and fabricating operations.[32] Thus embryonic changes in the sources of power and organization of production accompanied the expansion of processing activities and the emergence of a few producer-goods industries.

The conventional commercial pattern remained intact, for the largest proportion of establishments was interspersed among the commercial activities of the central area adjacent to the wharves, while others were scattered throughout the city (see Figure 3.1). The 1833 listing of manufactures both undercounted the total number of establishments and was biased in favor of the central area,[33] although the undercounting of some industries also occurred in the central area. On the basis of this listing, however, 47 percent of all establishments were located in approximately the same small central area that has been identified as a commercial core in 1800.[34] The few blocks west and north of this area contained an additional 20 percent. Under-enumerations of the three well-populated subareas of the city—Fell's Point, Old Town, and south Baltimore—may have inflated these percentages, but the centripetal nature of the pattern around the docks seems clear (Table 3.1 and Figure 3.1).

Although large and small shops of all kinds were interspersed in the central area and to a lesser extent scattered throughout the city, tendencies for localization also clearly existed. Brick making, quarrying, some food processing, and tanning were situated on the periphery of residential areas. A few foundries, machine works, planing mills, and chemical plants had begun to occupy the waterfront away from the central area in south Baltimore and Fell's Point. Shipbuilding was located there as well, although it is more difficult to document.[35] The city's central area also exhibited some internal patterning. The docks and immediately adjacent blocks contained ship-oriented activities, such as biscuit baking and sail making, and food processors, such as sugar refiners and distillers, as well as leather works and metalworking establishments. On the eastern edge of the central area along the low-lying plain of the Jones Falls, there was a similar variety of food processors and fabricators of leather, wood, and metal products. Moreover, the Falls area contained the largest number of steam-powered establishments. Directly west of the Jones Falls and a few blocks north of the docks lay the city's business area.[36] Although some heavy industry was present, the area was principally the focus of printers and consumer-serving artisan shops. Interspersed among the professional, financial, and retail businesses that stretched along Baltimore Street and the few blocks south toward the Merchants Exchange, tailors, hatters,

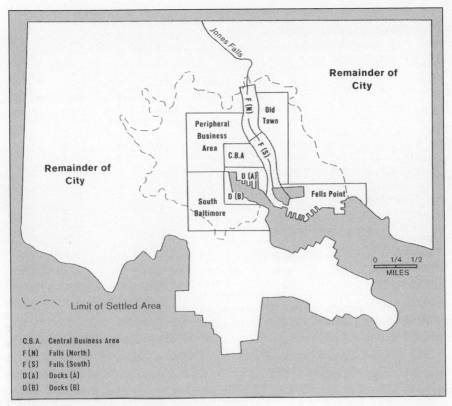

FIGURE 3.1. Areas in Baltimore, 1833.
Note: See Table 3.1 for data source and for number and type of manufacturing establishment by area.

silversmiths, watchmakers, and many other artisans produced high-value commodities, presumably for local consumption. At the western edge of the business area were leather and wood handicrafters, particularly saddle, harness, and coach makers, but the high-threshold artisan shops diminished sharply in number.

In 1833 the emergence of the new forms of production had not substantially altered the commercial pattern of intra-city industrial location. The location of large land-using or noxious industries on the periphery and waterfront and of steam-powered manufactures in the south Falls area provided the lineaments for later mid-century patterns, but the overwhelming proportion of activity and the most striking clustering occurred within the city's compact central area.

TABLE 3.2. Selected Characteristics of Manufacturing in Baltimore, 1860

Industry type	No. of establishments	% of total establishments	% of total value added by manufacturing	Total employment	% of employment	Female employment	% of female employment
Shipbuilding, repair, and related activities	42	3.7	5.7	605	3.8	0	0
Agricultural processing	122	10.6	19.2	973	6.0	0	0
Raw material processing	39	3.4	5.6	531	3.3	0	0
Construction materials	57	5.0	5.3	1,220	7.6	0	0
Metal processing and products	54	4.7	16.7	2,610	16.2	0	0
Clothiers	46	4.0	12.0	5,164	32.0	3,497	78.8
Artisan-dominated activities	786	68.5	35.4	5,017	31.1	942	21.2
All others	1	0	0	2	0	0	0
Total	1,147	99.9	99.9	16,122	100.0	4,439	100.0

Source: "Manufactures of the United States in 1860, Eighth Census," manuscript schedule, City and County of Baltimore (Maryland State Library, Annapolis).

The Transitional Pattern, 1860

Although annual newspaper reviews of business activity in the 1850s rarely acknowledged the existence of industry, the Eighth Census, in 1860, reported 1,147 manufacturing establishments with a total production value of $20 million and employment in excess of 16,000 persons (see Table 3.2).[37] The pattern of production that emerged from the census data depicted the transitional character of mid-century Baltimore. Traditional small artisan shops and commerce-oriented manufacturers coexisted with the newer organizational and technological forms of industry.

Artisan wares, processing of commercial commodities, and ship-related manufactures—the traditional pursuits—expanded with the overall growth of industry and amounted to approximately three-fifths of the total value added by manufacturing for the city. Although shipbuilders had fallen on hard times in the late 1850s, shipbuilding had generally been active since the 1830s.[38] In 1860 fifteen shipyards and related pursuits such as sail-making engaged more than 600 employees. Brewers, flour millers, slaughterers, sugar refiners, and oyster and fruit packers were prominent among agricultural processors. Copper, iron ore, cotton, wool, leather, and lumber were raw materials for nonagricultural processors.

In addition to those traditional kinds of activities, the small artisan shops continued to predominate, at least as measured by the total number of establishments. Approximately 55 percent of all manufacturing establishments employed fewer than five persons; 76 percent of them employed fewer than ten (see Table 3.3). The small shop was most conspicuous in local consumer-oriented industries. For example, 236 boot and shoe concerns averaged six persons per unit, while the 127 tobacconists and 41 metalware establishments averaged four employees.

Even though artisan shops accounted for a majority of industrial establishments, they employed less than one-fifth of the city's industrial workforce. The presence of many large firms, as measured by employment, represented an important change from patterns of the commercial city. In comparison to the few establishments with substantial workforces in the 1830s, nearly 54 percent of the city's manufacturing employees in 1860 worked for fifty-two firms that employed fifty or more persons. In all, 114 establishments employed twenty-five or more workers and provided employment for 67 percent of the industrial labor force.

This increase in the scale of industrial establishments in Baltimore resulted from the adoption of organizational changes such as the putting-out system and expanded workshop and from the widening application of

TABLE 3.3. Establishment Size for Selected Manufactures in Baltimore, 1860

Industry type	No. of establish-ments	% of establishments by no. of employees				
		0–9	10–19	20–49	50–99	100 or more
Iron foundries and machinery	28	21.4	0.0	39.3	25.0	14.3
Clothiers	45	12.3	22.2	35.6	6.7	22.2
Shipbuilding	15	46.7	6.7	20.0	13.3	13.3
Brick making	38	23.7	18.4	44.7	13.2	0.0
Cabinets and furniture	41	73.2	12.2	7.3	7.3	0.0
Tanning	22	77.3	13.6	9.0	0.0	0.0
Boots and shoes	236	86.8	8.5	4.7	0.0	0.0
Metalwares	41	90.2	9.8	0.0	0.0	0.0
Tobacconists	127	92.9	3.1	3.9	0.0	0.0
All steam-powered establishments	80	25.0	18.9	28.9	17.5	10.0
All city manufacturing establishments	1,147	76.1	9.9	9.4	2.8	1.8

Source: "Manufactures of the United States in 1860, Eighth Census," manuscript schedule, City and County of Baltimore (Maryland State Library, Annapolis).

steam power to the manufacturing process. The most striking change occurred in the ready-made clothing industry. The shift from tailored custom work to ready-made apparel permitted the use of many unskilled workers. Baltimore's forty-six clothiers employed almost one-third of the city's total industrial workforce, the largest producers reporting 800 and 1,250 workers. Moreover, two-thirds of the garment workers were female; the clothing industry employed 79 percent of all women employed in the city's industries (Table 3.2). It is impossible to determine what proportions and components of the clothing workforce were involved in the different forms of industrial organization, but it is clear that the clothiers used the putting-out system, the inside shop, and the relatively new factory system.[39] In addition to the clothing industry, a number of other activities had begun to adopt these organizational changes. Although small shops remained important, the makers of hats and shirts used the inside shop in combination with putting-out work, while oyster packers, carriage makers, and furniture producers preferred factory organization.

The rapid growth of Baltimore's iron foundries, machine works, and other metalworking manufactures further signified the industrial transition by their emphasis on producer goods, new transportation modes, and steam technology. Fifty-four iron and brass establishments, excluding the presumably local consumer-oriented metalware shops, employed 16.2 percent of the city's manufacturing labor force (Table 3.2). The Baltimore and Ohio Railroad's machine and repair works was the largest concern, with 990 workers, but even with the removal of that extreme figure the remaining shops still averaged 31.5 employees. Twenty-one of the twenty-eight iron foundries and machine shops, which had the largest workforces per unit, depended heavily on steam power. By 1860 eighty manufacturers of many different kinds of products reported the use of steam engines. In addition to machine works, steam power primarily permitted the large-scale processors of raw materials to locate within the city. Although there existed large-scale establishments that did not use steam, a majority of those manufacturers using steam employed twenty or more employees.

Thus mid-century organizational and technological changes in manufacturing substantially modified Baltimore's industrial structure. Conservative measures of the two most representative activities of that industrialization—the ready-made clothing and the metal industries—indicate that together they provided 28.9 percent and 48.5 percent of the city's total value added by manufacturing and industrial employment, respectively.[40] With the continued importance of shipbuilding, processing, and artisan production in the city, the question remains as to what extent the increasing prevalence of steam power, of large industrial establishments, and of new industries altered the location of manufacturing activity within the city.

Antecedent distributional characteristics associated with the commercial city were still in evidence, but centrifugal growth and functional clustering were also creating distinctive industrial districts within Baltimore. The concurrent existence of the old (particularly the scattering of shops throughout the city) and the new makes the transitional pattern more difficult to discern. Distributions of industrial activity are based on those establishments listed in the manuscript schedules of the census, for which it is possible to identify street addresses in the city directories or in a few other miscellaneous sources. In all, we were able to locate 726 establishments (66 percent of the total) employing 13,536 persons (84 percent of the total).[41] The data are aggregated into small grid cells of approximately three square blocks in order to compare 1833 data with those of 1860 because the earlier data are less locationally precise. Industrial districts are then defined on the bases of type of product and probable linkages among establishments, of spatial propinquity, of size of employment per establishment, and of mode of power (see Figure 3.2 and Table 3.4). Because grid cells are added together to form districts, there is some arbitrary bounding of industrial districts. For example, an obviously large and linked establishment may fall just outside a cell boundary, or in terms of the criteria listed above, a large portion of a cell might obviously not be part of a district. Similarity of activities in the districts clearly characterizes the areas, while the inclusion of unrelated activities undoubtedly captures the real nature of land use.

As in 1833, waterpower sites in the city's environs continued to support large plants engaged in the production of flour, textiles, iron goods, and paper. Those textile and iron works must be viewed as complementing Baltimore's growing clothing and iron industries. Within the municipal boundary, the familiar blend of noxious and building activities still bordered some residential areas. The removal of the livestock market only a few blocks from its previous site maintained the cluster of meatpackers on the far west side.[42] An extensive hair-and-bristle factory and a few tanyards were located near the market.[43] Brickyards were similarly concentrated on the city's southwestern edge, while breweries, flour mills, and other manufacturers were scattered around the remainder of the periphery.[44] Also in the traditional commercial city pattern, artisans located their shops in every ward. Although there were no outstanding clusters of such activities,

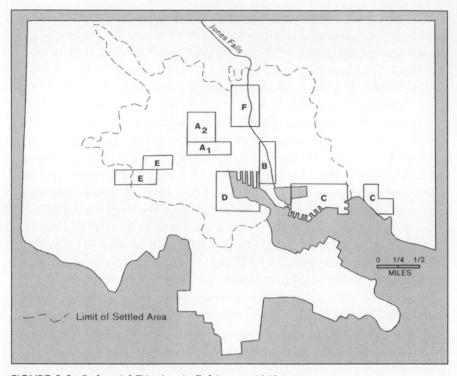

FIGURE 3.2. Industrial Districts in Baltimore, 1860.
Note: See Table 3.4 for data source and for manufacturing composition by industrial district (lettered areas on the map).

they did appear more frequently in south and west Baltimore and Fell's Point, where the largest number of German immigrants resided (see Figure 3.3).[45]

The decline of manufacturing in Baltimore's core represented one important change from 1833, a circumstance probably resulting from the heightened competition for central space by nonindustrial establishments. The number of manufacturers decreased by approximately 100, while the area's proportion of the city's total establishments declined from 47 to 15 percent. Only 16 percent of the city's industrial labor force worked in those central establishments. The docks and adjacent blocks contained the traditional food-processing, leather, and maritime-oriented manufacturers, but those establishments were both fewer in number than in 1833 and decidedly smaller-scale operations than their counterparts elsewhere in the city.[46] The business core around Baltimore Street, however, continued to attract printers and a variety of artisan shops.

The existence of a large boot and shoe manufacturer and a large cloth-ier on the western edge of the central business area marked an abrupt transition to an adjacent warehouse and workshop district.[47] Even as late as 1860, a Baltimore newspaper noted the recent rapid transformation of the blocks along this edge of the original business core to warehousing.[48] Here, among warehouses, small businesses, and residences for an area approximating thirty blocks, was the greatest concentration of manufac-turing in the city, with eighty-nine establishments employing one-quarter of the industrial workforce (Table 3.4). The ready-made clothing indus-try overwhelmingly predominated, but other activities contributed to the warehouse-workshop character of the district. The location in 1833 of dry goods importers, textile warehouses, and hardware importers around a focus of Baltimore and Howard streets anticipated that mid-century devel-opment.[49] In 1860, ten of the city's largest clothiers were in District A1 (Figure 3.2). They employed approximately 4,000 persons, most of them women (Figure 3.3). The putting-out system undoubtedly dispersed some of that workforce, but the use of multistory buildings by some clothiers signified the daily concentration of garment workers.[50] Another sixteen clothiers and clothing-related activities with more than 2,000 workers (50 percent of them female) existed a few blocks north within the general

TABLE 3.4. Composition of Industrial Districts in Baltimore, 1860

Industrial employment districts	Total no. of establish-ments	Total employ-ment	No. of steam-powered establish-ments	Employ-ment in steam-powered establish-ments	Total female employ-ment	Female employ-ment as % of total employ-ment
D	75	552	4	130	47	8.5
C	54	1,204	12	734	126	10.5
B	38	624	9	438	14	2.2
F	31	735	10	378	1	0.0
A1	16	4,090	0	0	2,885	70.5
A2	73	732	2	93	168	23.0
E	14	1,513	4	1,350	2	0.0
Total	301	9,450	41	3,123	3,243	34.3
% of all located industries	41.2	69.8	60.3	85.9	82.4	
% of all industries	36.1	58.6	50.1	81.3	73.1	

Source: "Manufactures of the United States in 1860, Eighth Census," manuscript schedule, City and County of Baltimore (Maryland State Library, Annapolis).

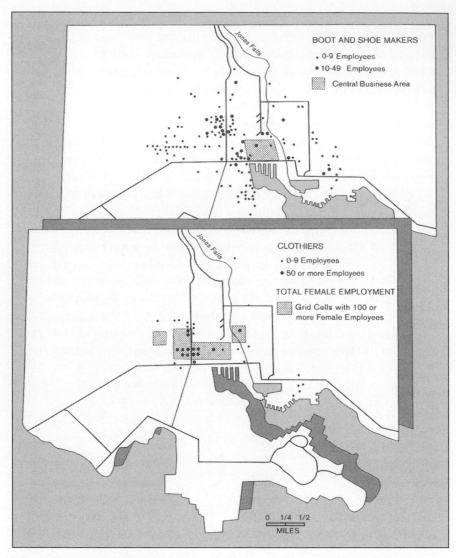

FIGURE 3.3. Boot and Shoe Makers, Clothiers, and Female Industrial Employment in Baltimore, 1860.
Note: See Table 3.4 for data source.

workshop-warehouse area of District A2 (Figure 3.2). Although small artisan shops were most numerous, the city's medium-sized workshops (more
than ten employees) in leather and wood products were conspicuous here
as well. Only two establishments reported the use of steam power. Thus,
handicraft production, large workforces, and high proportions of female

workers characterized manufacturing in this area. As in other eastern sea-ports at this time, workshops developed in inexpensive quarters proximate to merchants, wharves, and unskilled labor.[51] Although the employment of almost 5,000 manufacturing workers in the warehouse-workshop dis-trict added to the centripetal perception of the city, in comparison to ear-lier patterns it reflected a distinct localization separate from, though adja-cent to, the maritime and business cores.

On the eastern edge of the central business area in the low-lying land of the Jones Falls, there was more continuity with the past. Along with numerous artisan shops, more than a dozen food and material processors, many dependent on steam power, employed several hundred male work-ers in District B (Figure 3.2). Sugar refineries, breweries, saw mills, plan-ing mills, foundries, and machine shops created a composition of activi-ties similar to that in 1833, but the total number of steam plants declined. The vulnerability of the south Falls area to sudden devastating floods might have discouraged growth, and, as the area had long been settled, lit-tle open space remained for additional industrial development.[52]

The clustering of shipbuilding activities along the waterfront of Fell's Point in District C (Figures 3.2 and 3.4) was another link with the com-mercial city. Besides the half-dozen shipyards, many other manufacturers made products such as sails, oilcloth, brass fittings, iron plating, lumber, and steam engines for the shipbuilding industry. Together, the twenty-four ship-oriented manufacturers employed almost 900 workers, pre-dominantly male. The remaining 300 industrial workers of this cluster were found in a few large miscellaneous activities and in numerous local consumer-serving shops.

The three remaining districts had the attractions of available land and regional transportation facilities. District D included the waterfront area of south Baltimore (Figures 3.2 and 3.4) and contained a variety of mar-itime-oriented industries in 1860, but the district was still relatively small, offering employment to fewer than 600 persons. Shipyards, foundries, machine shops, and food- and material-processing facilities dominated the shoreline, while a number of artisan shops, mostly consumer-serving, were located nearby in the south Baltimore retail area. Approximately one mile to the west, however, was an area of foundries and machine shops, Dis-trict E (Figures 3.2 and 3.4), that formed the city's second-largest cluster of industrial workers. The Baltimore and Ohio Railroad shops employed 990 workers, and five other large metalworking establishments were nearby.[53] With the addition of two brickyards and a woodenware plant, more than 1,500 persons worked in the fourteen establishments of this far west-side cluster. Presumably in a similar manner, the low land and the

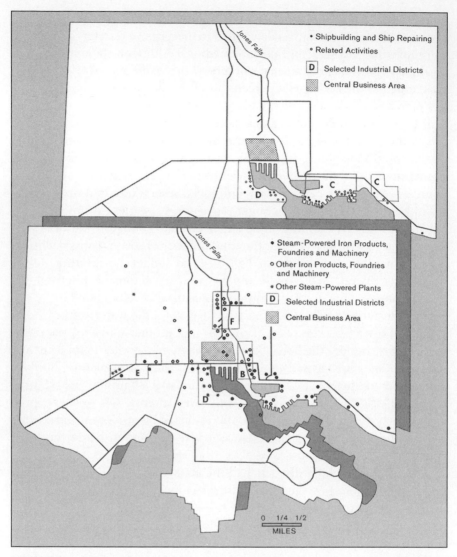

FIGURE 3.4. Shipbuilding, Ship Repairing, and Related Activities, and Iron Products and Steam-Powered Plants in Baltimore, 1860.
Note: See Table 3.4 for data source and for location and employment size of steam-powered plants by industrial district (lettered areas on the map).

railroad facilities of the Baltimore and Susquehanna Railroad in the northern basin of the Jones Falls, District F (Figures 3.2 and 3.4), attracted several large plants. A few processing mills had long used the waterpower of the Falls in this area, but by 1860 steam engines provided the power for tanneries, distilleries, and a flour mill. In addition, a textile factory, three marble-cutting plants, and several metalworking industries employed large workforces. Altogether more than 700 persons worked in the manufacturing establishments of the north Falls basin.

In total, the trends toward localization by workshops and factories resulted in the development of six districts of industrial activity, half of which reflected earlier locational tendencies and half of which represented new departures. These six districts contained 40 percent of the city's industrial establishments and 70 percent of its industrial workforce in 1860. They encompassed a majority of the city's large-scale industrial operations. The districts were differentiated by product, technology, and labor force characteristics, as indicated by the different craft requirements and the female composition of the warehouse-workshop area.[54]

Implications for Mid-Century Structure

The substantial developments of ready-made garment production and steam-powered metalworking activities, as well as the inclusion of females in the industrial labor force, indicated that industrialization was well under way in mid-century Baltimore. Detailed analysis of the locations of manufacturing establishments in 1833 and 1860 show the emergence of six industrial districts that resulted from centrifugal and localizing tendencies among large firms. Those districts were composed of complementary and similar manufacturing activities that together in each district employed hundreds, and in one instance thousands, of workers. The industrial districts existed simultaneously with the traditional commercial pattern of small artisan shops and processing establishments that were both focused on the central waterfront and widely scattered throughout the city. Perhaps the complexity of concomitant, but differing, distributions of the various industrial types and productive modes has obscured the identification of the evolving pattern in mid-century cities. Documentation of the functional differentiation and spatial separation of Baltimore's industrial districts not only affirms similar developments in Philadelphia but, through the identification of the composition of such districts, has implications for our understanding of the social geography of cities.

The central business area of Baltimore, including the growing warehouse-workshop district, remained the predominant locus of diversified employment. The concentration of large-scale handicraft producers and female workers distinguished the warehouse-workshop district from the evolving downtown and from the other industrial districts. Four of the six districts were clearly not in or adjacent to the city's central area. In at least two instances, they constituted an initial wave of industrial suburbanization. Located on the edge of the city's built-up area, those districts were characterized by their distinctive industrial structures; however, each also contained unrelated artisan shops and nonindustrial businesses. There is only scattered evidence from Baltimore and other cities about employee housing built by employers on the urban periphery, the prevailing pedestrian mode of getting to work, and the subcity organization or orientation of many retail, municipal, and community services. Nevertheless, on the basis of that evidence it is reasonable to speculate that the transitional city was evolving a fragmented or cellular structure.[55] In the context of pedestrian movement, immense population growth, and peripheral expansion, mid-century industrial localization may have contributed to the development of subcity areas within which people found employment, essential services, and social amenities. Interestingly, the industrial districts documented here coincided approximately with locally perceived areas of the city.[56]

Although there is evidence of small prestigious residential areas, middle-income suburbs, and ethnic or racial enclaves, the social geography of the large North American transitional city remains largely unspecified. High population densities, scarce housing, and an apparent lack of strong segregation along economic, ethnic, or racial lines have left an impression of residential heterogeneity.[57] Social and functional elements of a commercial society may partially underlie the persistence of residential heterogeneity, but the peripheral development of substantial and diverse industrial employment centers, along with the concomitant growth of supporting service activities, would also have led to the growth of heterogeneous residential areas and yet at the same time would have given some order to social space.[58] Rather than an indiscriminate view of residential location based on housing scarcity, people may have recognized and functioned within subcity areas of distinctive employment opportunities and social networks. Only the later lengthening of employment linkage for a much broader proportion of the urban populace, and the massive industrial recruitment of ethnic groups, would result in residential patterning by ethnicity and race.

4

Model City?

Industry and Urban Structure in Chicago

MARY BETH PUDUP

Chicago is the *known* city.

—Richard Wright

hicago has long served as a model, if not *the* model, for understanding urban development and spatial structure in the United
States. Its location in the middle distance between east and west,
coupled with its sheer size and abiding economic and social
diversity, has encouraged an unwitting acceptance that Chicago
is an ideal typical U.S. city. History books with popular and
scholarly inflections routinely proclaim Chicago the "most
American" of American cities. But Chicago's status as a model
city rests on more than obvious facts of its geography or unques-
tioned assumptions about its history. Since early in this century,
Chicago's historical geography has inspired scholarship whose
methods and findings have made the city larger than life.

The "Chicago School of Sociology," associated with University of Chicago practitioners Robert Park, Louis Wirth, and
Ernest Burgess, played a central role in marking Chicago as paradigmatic of urban process and form, captured in the signal
and simply titled collection *The City*.[1] Although the early twentieth century witnessed analogous social studies of other U.S.
industrial cities, notably the New York and Pittsburgh surveys,
the social studies of Chicago achieved an elevated status.[2] For

generations after, really up to the present, scholars in and outside of geography continued to reckon with this legacy.[3] Precisely why the Chicago School exerted such influence remains unclear. The prominence of associated scholars is obviously an important factor. Also relevant, but more elusive, is the willingness of consumers of Chicago social studies, that is, the wider intellectual community, to accept the frameworks and findings of the Chicago School as the foundations of a modern urban epistemology.[4]

A key supposition in the Chicago School's conceptual framework is the notion that cities gradually grow outward from a central core area. Incremental growth precipitates the emergence of tidy zones of similar land use (offices versus homes, warehouses versus factories) and similar social composition (upper class versus working class, European immigrants versus blacks).[5] This essay examines aspects of Chicago's historical economic geography and suggests the extent to which the city's spatial development departed from these maxims. The essay emphasizes how, from the very outset, the nascent metropolitan region grew in several locations simultaneously. Land subdivision throughout the "periphery" commenced early on, as did the establishment of outlying settlements. The spread of Chicago's contiguous settlement, coupled with a series of municipal annexations, negated the spatial separation of these early "suburbs"; this effacement, in turn, has implications for models of urban land use.

Exclusive residential suburbs such as Evanston (to the north), Riverside (to the east), and Hyde Park (to the south) contributed significantly to Chicago's multinucleated spatial development during the nineteenth century. This much is well known.[6] Moreover, the multifarious role of residential real estate developers in structuring Chicago's urban landscape, as in cities across the United States, has been well documented.[7] Less acknowledged, however, is how industrialization contributed to Chicago's multicentered metropolitan formation. Likewise, that the localization of industry within Chicago commonly reflected the handiwork of real estate promoters, rather than the atomistic behavior of industrialists, is too rarely recognized.

The study of industrial location, whether at the international, national, regional, or local (intra-urban) level and regardless of theoretical foundation, has a long tradition of ignoring micro-level conditions governing how investment decisions take their places on the landscape. Instead, the emphasis of industrial location scholarship has consistently been on firm-level behavior and, more recently, on how macroeconomic conditions structure investment and location decisions.[8] The historical geographical space between a firm's decision to invest in a factory in an individual city or town and the factory's operation in the actual place it was constructed is rarely explored in location literature. Although written seventy years ago,

Mitchell and Jucius's summary still rings true: "students of plant location have usually been so much engrossed in attempts to analyze the various economic and geographic factors, which presumably determine the points to which industries go, that it is possible that too little attention has been given to what may be termed the 'institutional' phases of the problem."[9] The apparent assumption is that nitty-gritty details of location—why along this rather that railroad? on this parcel of land rather than one three miles away?—are outside the purview of locational analysis. Such micro-level inquiry is characteristically left to the typically more place-based analysis of historical and urban geography. But if we take seriously the notion of geographical industrialization—that industries build cities—then we must fully interrogate this space from both sides of the dialectic: the location of industry and the biography of place.[10] We must merge analysis of the steel industry's location in Chicago, for example, with explanation of why the industry became concentrated in the southern reaches of the urban region.

This chapter seeks to contribute to the mutual re-mapping of industrial geography and urban historical geography by exploring not just where but, crucially, *how* industry came to occupy its place(s) in Chicago's emerging metropolitan landscape. Throughout the chapter, I pay particular attention to developments in the southern suburban region of Chicago during the nineteenth and early twentieth centuries. Using the case of the Calumet district, I first discuss how Chicago's commerce in natural resource commodities shaped long-term patterns of outlying settlement, industrialization, and industrial location. Considered next is the role of land developers in bringing suburban industrial real estate onto the market, as in the cases of Grand Crossing, Harvey, and the South Branch district. The third section is a brief interlude about the railroads as key players in shaping the urban industrial landscape. The essay then examines Chicago's Union Stock Yards, Packingtown, and the neighboring "Back of the Yards" working-class community as vivid exemplars of how the interplay among real estate promotion, railroads, and industrial location constructed the city. Finally, examination of Chicago's nineteenth-century development provides the basis for a concluding appraisal of the Chicago School's model of urban form and its legacy.

Nature's Metropolis: From Farm Town to Steel Town

Historians of early Chicago invariably attribute the city's economic growth to its commerce in agricultural and other natural resources.[11] As the farm population in northern Illinois increased during the late 1830s and early

1840s, Chicago began to outdistance other western cities, especially its chief competitor, St. Louis, as the market destination of choice for mid-western farmers. In Homer Hoyt's words, "There the highest prices in the West were paid for wheat that was shipped to the East on lake vessels and there the lowest prices were charged for salt, cook stoves, lumber, and other farm necessities."[12] Hoyt claims that lumber was one of the most important commodities farmers hauled back home from Chicago, and, indeed, receipts of white pine lumber from Michigan forests soared during the 1840s, from 7.5 million board feet in 1843 to 32 million in 1847.[13] Although conditions of land transportation posed considerable difficulty during the 1840s, a large flow of people and goods passed through Chicago. The expense, not to mention the inconvenience, of hauling produce and goods over bad roads to and from Chicago was eliminated in 1848 with the opening of the Illinois and Michigan Canal.[14]

Agricultural commerce stimulated some of the first settlements on the periphery of the incipient metropolis. Fellman studied land subdivision outside the built-up area of Chicago, what he termed "pre-building growth patterns," and asserted that "the several isolated, outlying blocks of land subdivided before 1855 are, with few exceptions, correlated with main trails" (Indian traces and wagon roads) that were corridors of farm trade.[15] The significance of these transportation arteries in shaping Chicago's spatial structure is often lost amid the fanfare surrounding the advent of the iron horse. While land subdivision took place along trails leading into Chicago from all three directions—south, north, and west—the area to the south and southwest of the city appears to have experienced the earliest and fastest development. These areas endured as the principal loci of Chicago's industrialization. As Hoyt explains, "the movement of wagons into the south division of the city was from the first by far the greatest," because the main trails, the "Chicago Road," Vincennes trail, and "Road to Widow Brown's," entered from already settled areas to the east, south, and southwest, respectively.[16] The direction of these important trails and the location of early outlying settlements correlate with the emergence of the area adjacent to the south bank of the Chicago River as Chicago's central business district.[17]

As commerce between city and hinterland increased, small market towns emerged near stopping points along the connecting roads. In time, these became nodes around which the city would grow. This can be seen in the case of the South Chicago area of Roseland along one such access route.[18] Dutch emigrant farmers first settled the area during the 1840s and Roseland quickly became a stopping place for farmers making their way to the Chicago produce markets eleven miles north on South Water Street. In

1850 a tavern called Eleven Mile House was built there as a refreshment stand and gathering place. Throughout the middle decades of the nineteenth century Roseland's farming population continued to increase, and by 1880 numbered 700. By that date the area also boasted a school, post office, and several churches and enjoyed two railroad connections at nearby Kensington. The area immediately south of Roseland underwent a dramatic transformation beginning in 1881, when George Pullman decided to construct his mammoth sleeping car works and model town along Lake Calumet.[19] Roseland itself changed as a result of the more than 1,500 dwelling units constructed by the end of the century to house workers crowding into South Chicago's burgeoning factories. By 1900 Roseland boasted at least five steel companies, three lumberyards, and a variety of other concerns.[20]

Indeed, by the turn of the century, the Calumet district of South Chicago, which embraces Roseland, had become a principal center of Chicago's industrial growth. Industrialization in the Calumet district traces its origins to farsighted land investments and infrastructure improvements immediately following the Civil War. Even before the war, the Calumet district was recognized as a promising site for a city—more so, in fact, than the site of Chicago. The Calumet is a far larger stream than the Chicago River and offered better natural harbor possibilities at its confluence with Lake Michigan and lakes immediately inland. When the two rivers were surveyed in the 1830s for the purpose of siting the Illinois and Michigan Canal, technical reports favored the Calumet outlet. But, as Everett Chamberlin noted dryly in 1874, "engineering did not decide the point."[21] The central city land and business investments of Chicago's incipient elite, who were the most avid canal promoters, precluded any serious consideration of the Calumet portage as the future central business district.

The promise of the Calumet was not forgotten, however. During the Civil War, Colonel James Bowen began making plans to turn the natural advantages of the district into comparative advantages. In 1868 Bowen assembled a group of investors and purchased large tracts of land adjacent to the Calumet harbor.[22] The following year the group obtained a charter from the Illinois legislature as the Calumet and Chicago Canal and Dock Company. The charter spelled out plans to construct and operate a canal linking the Calumet River to either the south branch of the Chicago River or the Illinois and Michigan Canal, and to build and operate docks, dry docks, shipyards, warehouses, piers, and all the other accouterments of a working port. Additional land purchases brought the company's acreage up to about 6,000 and included properties suitable for different land uses.

Investors did not leave the success of the Calumet and Chicago Canal and Dock Company to chance. They spent more than a quarter-million dollars on improvements designed to enhance the value of the property and attract other investors. The company financed a macadamized road between central Chicago and the Calumet River, local street grading, and river dredging, and constructed a hotel and other commercial buildings. A long list of railroads had direct connections in the district; an even longer list had indirect access through the Chicago Belt Railroad.

The long-term strategy of the Canal and Dock Co. was to expand the city southward to the boundaries of its property. Actual development did not wait for this to happen. By 1873 the Calumet district boasted several manufacturing establishments representing the range of Chicago's industries. These included a woolen mill, a steam forge works, a gristmill machinery company, a match factory, and steel works. The company reportedly extended "every favor and every aid possible" to manufacturers who located on its land, and stood ready to provide "favorable inducements" to others that would follow. In his *Chicago and Its Suburbs*, a promotional tool for real estate development throughout the Chicago region, Everett Chamberlin lauded improvement efforts in the Calumet.[23] He emphasized that establishment of these industries would promote not only further industrialization of the Calumet but also investments in large tracts for industry even further south of the city. By the early twentieth century, the Calumet district had become the center of Chicago's vast iron and steel industry, though, as Hoyt points out, "grain and lumber were scarcely less important" in the Calumet's remarkable growth.[24] Traffic on the Calumet River, once shunned in favor of the Chicago River portage, was five times as great as on its rival. The region's development spilled over state boundaries, too, with the founding of port and manufacturing sites at Indiana Harbor and Gary in 1903 and 1906, respectively.

The full elaboration of industrial development in the Calumet district of South Chicago did not "result" from the area's early proximity to southerly overland routes important to early Chicago's agricultural commerce. Yet proximity does figure in the selective growth of this area within the urban region. Similar processes of selective growth account for the emergence of other settlements along trails associated with the city's overland livestock trade.

As with the grain trade, the city early on became a market destination for hogs and cattle. Providing stockyards for animals and services and amusements for livestock drivers stimulated peripheral settlements where "more than just animals and money changed hands."[25] A case in point was

Brighton, an area southwest of Chicago's original Fort Dearborn settlement. By 1850 two roads near Brighton, Blue Island (today's Western Avenue) and Archer, had become important livestock trails leading from northeastern Illinois and northwestern Indiana into Chicago.[26] Drovers using Blue Island and Archer roads typically herded their cattle past Brighton to stockyards that had been constructed further along at Bridgeport. Recognizing a missed opportunity, several enterprising Brighton businessmen and landowners built stockyards and a hotel at Brighton in a successful effort to divert the livestock traffic. Hoping to encourage settlement, they subdivided surrounding land and invested in road improvements. Infamous Chicago mayor and real estate speculator "Long John" Wentworth constructed a racetrack as an additional attraction. By the early 1850s Brighton was well on its way to becoming a concentrated settlement.

Brighton did not develop into Chicago's central livestock market, however, as investors had hoped. After initial success, the settlement's subsequent history was more prosaic. Other stockyards located along new railroad lines soon out-competed the Brighton market. The final death knell was sounded when the Union Stock Yards opened in 1865. A few industries were located in Brighton during the 1870s, and the Santa Fe Railroad built its Corwith Yards there in 1887.[27] But instead of becoming a center for industrial development in its own right, the fortunes of Brighton Park, as the residential community became known, were tied to industries in nearby areas.

Brighton Park's significant growth took place after it was annexed to the city of Chicago in 1889 and it became a working-class neighborhood populated by a variety of immigrant nationalities.[28] "With about 1,000 dwelling units in 1899, there would be over 6,000 by 1920, and over 10,000 by the 1929 crash," according to Keating's *Building Chicago*.[29] Of all the agricultural commerce towns that had dotted Chicago's periphery, Brighton enjoyed the closest proximity to the city center and as a result grew rapidly, with extension of contiguous urban settlement and industrial development. Residents continued to be attracted to Brighton Park by the industry developing around the Santa Fe yards, by the better housing available in the neighborhood, and by the area's excellent streetcar connections with both the central city and Union Stock Yards district. The Central Manufacturing District, opened in 1905, provided employment to Brighton residents, as did the Crane Manufacturing Company, which relocated nearby in 1915. Meanwhile, the Kenwood Manufacturing district, on the southern edge of Brighton, opened for business in 1915.[30] By 1930 Brighton Park had been fully woven into Chicago's urban industrial fabric.

Its continuing growth at the turn of the century illustrates how working-class residential areas grew up simultaneously with industry.

Antebellum peripheral settlements like Brighton and Roseland originated by servicing the city's commercial economy through agricultural processing, trade, taverns, and hotels. At first glance, such settlements appear to be natural outcomes of the dynamic interaction between Chicago and its widening resource hinterland. But even in these early peripheral settlements was evident the visible hand of real estate development that left its prints everywhere on Chicago's urban landscape.

Man-Made City: Property Developers as Industrial Locators

Before a single bushel of wheat or herd of livestock ever entered the city, Chicago was in the business of buying and selling land. "In these early days," as Mayer and Wade put it, "Chicago seemed less a community than a real estate lottery."[31] The city's incorporation into the national economy took place during the run-up in real estate values preceding the panic of 1837. The situation in what was then still a frontier outpost apparently was no less frenzied than elsewhere in the nation. Swampland of questionable value changed hands several times over, each time at a price considered unimaginable. Hoyt vividly describes conditions along the south bank of the Chicago River in 1835, when land values in Chicago were on the rise but had not yet reached their peak, as they would the following year.

> The Tremont House lot at the corner of Lake and Dearborn (80 by 180 feet) that could have been bought for a cord of wood, a pair of boots, and a barrel of whiskey in 1831, 1832, and 1833, respectively, was now valued at a sum of money that would fill a warehouse with such commodities. Dole's corner, at Dearborn and south Water street (80 by 180 feet), was sold for $9,000 in March of 1835 and $25,000 in December. One hundred thousand dollars was offered and refused for Hogan's block at 272 Lake Street.[32]

While much of the speculation involved land in the central business district, land fever engulfed the entire urban region, especially suburban areas along the lakeshore and north and south branches of the Chicago River, along which industries were congregating. Development of Chicago land was postponed by the deep financial depression that did not abate until 1843. Unlike many other western places, where real estate speculation left nothing but a trail of paper, Chicago weathered the panic thanks to the resolve of its bourgeoisie, who shared common interests in advancing their own fortunes by advancing those of the city. "From its settlement in the

1830s until the 1860s Chicago was dominated by an elite of exceptional energy and uniformity in background," according to Gerald Suttles. "These men, most of them from small-town New England, engaged in real estate development, merchandising, and especially the transshipment of raw or partially processed goods."[33] The elite proved particularly adept at securing public financing for the great public works projects that were the foundation of the modern city, such as harbor improvements, extension of the grid pattern, construction of the Illinois and Michigan Canal, and raising of the city above the swampy levels of lake and river.

Historical evidence suggests that Chicago's elite exercised considerable skill in packaging land development projects. Developers of residential areas, of which there were many in the city, used an array of "service improvements, like water lines, sewer pipes, and gas connections to attract a homogeneous clientele to their subdivisions on the outskirts of Chicago."[34] Suburban industrial land development followed similar rules of providing improvements for manufacturers, including rail sidings, paved streets, and even factory design and financial assistance. Like the developers of residential subdivisions, industrial land promoters enjoyed great success in Chicago. Land for industry was brought onto the market by lone investors, partnerships, group investment schemes, and eventually, in the early twentieth century, by railroad corporations in the guise of planned industrial districts.

Chicago real estate promoter Paul Cornell is perhaps best known as the developer of the exclusive preserve of Hyde Park, one of Chicago's oldest residential suburbs. The heart of Cornell's development was a 300-acre lakeside parcel he acquired in 1855, next to which he constructed the Hyde Park House hotel. Cornell also sold a tract of his land to the Illinois Central Railroad and entered into a contract for regular stops at his hotel.[35] Having put together favorably situated land with scheduled rail service, Cornell exhorted wealthy Chicagoans "to create a 'resort' and a 'retreat' from the congestion of the city" in a highly effective advertising campaign.[36] This vision of Hyde Park was fulfilled only after the close of the Civil War, which dampened real estate development throughout Chicago and the rest of the nation.

Cornell's other major suburban Chicago venture, Grand Crossing, lay southwest of Hyde Park and took its name from the fact that the land straddled three railroad lines. Cornell acquired this land also in 1855, but from the start he entertained a vision very different from Hyde Park: a manufacturing town that would benefit from the concentration of railroad connections. After the wartime interruption, Cornell filed a plan for subdividing the Grand Crossing property in 1871 and began issuing promotional pamphlets trumpeting his industrial vision. Within two years 150 trains passed through Grand Crossing every day, and, because of treacherous grade crossings, all were legally required to stop. Cornell's enterprise was so successful that by 1876 nearly a dozen manufacturing establishments had located at Grand Crossing, including a furniture company, a sewing machine company, and a rolling mill. "The settlement at Grand Crossing began to resemble Cornell's dream of a manufacturing town," in Keating's words.[37] After this initial growth, Grand Crossing continued to prosper as a working-class industrial community distinct from Chicago. But by the turn of the century, the extension of streetcars and cable cars to the vicinity, coupled with the deepening industrialization of southern Chicago and the expansion of contiguous settlement, strengthened the connections between Grand Crossing and surrounding communities. By 1930, roughly the same time as Brighton Park, Grand Crossing had become fully integrated into Chicago.[38]

The development of Harvey, a manufacturing suburb about eighteen miles south of the Loop, illustrates how industrial real estate operators often included specific provisions for worker housing in their plans.[39] In 1889 lumber merchant Thomas Harvey bought an already subdivided parcel known as Southlawn along the Calumet River and formed the Harvey Land Association for the purpose of constructing a self-contained "manufacturing suburb." Harvey and his religious partners sought to instill the spirit of temperance in prospective residents and therefore banned the sale of alcohol within town limits. More generally, Harvey incorporated the range of modern conveniences such as electric lighting, sanitary hookups, graded schools, and a trolley system to transport workers to their jobs along landscaped boulevards. Financing was made available to blue-collar workers to enable them to purchase land and a cottage for about $1,000. By all accounts, the town of Harvey was an instant success. A mere three years after its founding, ten major industries and 5,000 residents called it home.[40]

Other areas in southern Chicago that became major industrial centers originated in speculative ventures that united real estate subdivision with the extension of railroad lines. The South Branch manufacturing area was "almost exclusively" the enterprise of Samuel J. Walker, who purchased a full one-and-a-half miles of river frontage land in 1854 for the express purpose of siting industry.[41] The timing suggests that Walker was investing with a long-term view of the city's industrial areas eventually extending to his tract. Walker's land embraced almost 1,500 acres, which extended from the river to beyond the Illinois and Michigan Canal. His improvements included dredging an initial six slips for docks, which added 20,000 feet of frontage. His South Branch district also contained 10,400 feet along on the Illinois and Michigan Canal. All told, the subdivision boasted 45,000 feet of dock frontage, or more than eight lineal miles. In addition to excellent waterfront capacities, Walker invested in a private railway to serve the head of each slip and provide connections between them. Five railroads traversed the South Branch district, and the nearby Union Stock Yards provided switching facilities to every other road entering the city.

For many years Walker struggled to make a go of the South Branch. During the 1860s he entered negotiations with several manufacturers, including the McCormicks, of reaper fame, to buy land, and built large new factories—but these came to naught. Walker's South Branch fortunes finally turned around in the early 1870s, stimulated by land-use changes and quickening industrialization that came in the wake of the great Chicago fire of 1871.[42] Rosen claims that the South Branch industrial area took off when the McCormick Brothers finally relocated their massive reaper works there in 1873, after the fire destroyed the company's factory on the north bank of the Chicago River. The new reaper works was the largest farm machinery factory in the world, containing several buildings spread over twenty-three acres, including two furnaces capable of smelting thirty tons of pig iron per day and a giant 300-horsepower, sixteen-ton steam engine that powered all the machinery in the other factory buildings. "Newspaper reports show that the McCormick Brothers' decision . . . triggered the take-off of the area," according to Christine Rosen, "by stimulating interest in the region on the part of other manufacturers (and the press) and by spurring the expansion of the area's transportation facilities, which itself stimulated further interest."[43]

The South Branch district was an immensely successful industrial real estate investment. By the end of 1873, "several of the factories were among the largest and most important in the city. Built to take advantage of unprecedented economies of scale, many covered several acres of land and were engaged in the heaviest of heavy iron and steel manufacturing."[44] In addition to the reaper works, other notable firms included Wells, French & Co. (railroad bridge and car works), F. E. Canada & Co. (bridge and car works), Columbian Iron Works, Union Rolling Mill Co., and Joliet Rolling Mill Co. The South Branch also contained thirty brickyards and a furniture factory. Together these factories employed more than 3,000 workers.

The development of South Branch, like that of Grand Crossing and the Calumet district, supports Wrigley's contention that "in the Chicago area,

... the scientific planning of industrial land use in the United States ... reached its greatest development."[45] The successes of early industrial land developers typically depended on a combination of shrewdness and determination to steer investment, luck in owning land in an existing or planned transportation corridor, and a healthy measure of perseverance in waiting out wars, disasters, and business cycles. Samuel J. Walker's investment in land along the South Branch easily exemplifies all of these qualities. During the nineteenth century, individuals and syndicates were the main promoters of industrial land development. Only in the early twentieth century did corporately organized industrial districts assume a significant role in shaping Chicago's urban landscape, with railroad companies taking the lead.

The Central Manufacturing District (CMD) is credited with being the first such organized industrial district in Chicago, indeed the nation. The district was organized in 1890 by the Chicago Junction Railways and the Union Stock Yards Company to develop parcels of land in the area directly north of the stockyards. The site contained ungraded vacant land as well as a vast expanse of lumberyards whose glory days had passed. Given the preponderance of manufacturing already in the area, the CMD company wisely sensed the possibility of redeveloping the site for new industrial purposes. Not incidentally, another rationale for organizing the CMD was to provide the railroad with additional carrying trade. The Stock Yards and Junction Railway companies together operated excellent yet highly localized facilities and services developed to move livestock and perishable packinghouse products quickly between the Union Stock Yards area and trunk line carriers that in turn moved the goods to distant markets. Organizers of the CMD realized that these facilities and services "could be extended to serve equally efficiently a territory of much greater extent. It was only necessary to find and create this new realm of usefulness."[46] Thus, from the outset, while the CMD had its sights set on industrial land promotion, its long-range goal was to enlarge the carrying trade of its founding companies.

The better use of existing railroad facilities appears to be a common thread in the histories of the several other organized industrial districts in Chicago, including those located in suburban areas. For example, the Clearing Industrial district was organized in the early 1890s by the Chicago Great Western Railroad for the purpose of building and operating outside the southwestern edge of the city a large clearing yard to facilitate the movement of freight cars among railroads entering Chicago. This much-needed facility promised to partially relieve chronic traffic problems arising from the unregulated expansion of the city's rail network. For a variety of reasons, the classification yard plan stalled and Clearing officials turned to developing part of the land for manufacturing purposes. The classification yard plan eventually become reality in 1912, when the Belt Railway of Chicago, a circumferential connecting line jointly owned by thirteen major railroads, purchased the Clearing district's railroad facilities and quickly built vast new classification yards. By virtue of its proximity to the new yards, the original Clearing district profited handsomely by being able to offer door-to-door connections with industries located along all the railroads entering the city.

Chicago's organized industrial districts can be thought of as precursors to the suburban industrial parks that became commonplace in the United States during the post–World War II era. Their owners vigorously promoted the advantages of locating industry within their boundaries. Moreover, because both the Central and Clearing districts were composed of several different tracts located throughout southern quadrants of the city, promoters could tailor their offerings to meet specific industrial needs.[47] There were six different tracts in the CMD, for example, including one in the Calumet region that was "evidently acquired to offer sites for heavy industrial purposes."[48] When soliciting out-of-town industries, promoters could enlist the Chicago Association of Commerce and Industry to extol the virtues of a Chicago location. Over the course of the twentieth century the districts offered an expanding array of services to industrial prospects beyond the basics of rail sidings, streets, and public utilities. Small businesses in particular took advantage of expanded services like industrial building design, finance, and construction. The districts even sponsored organized social clubs and sporting events and established committees charged with studying municipal problems like taxation and traffic. In Chicago, the organized districts did not so much innovate as they codified a range of practices pioneered by early industrial land developers.

Railroads, Real Estate, and Urban Space

How can we understand the agency of railroads in the organization of industrial districts and as shapers of Chicago's urban industrial space? Historians and geographers alike have been virtually unanimous in explaining nineteenth-century Chicago's growth in terms of transportation improvements, especially expansion of the American rail network, in which the city assumed a central position.[49] Defined technically as infrastructure, railroads are seen to bear people and goods across space, reducing the time and cost of overcoming distances and expanding market areas. Railroads facilitate the expansion of trade, which in turn stimulates commodity production—agriculture in the country and industry in the city. Railroads were the iron ties that bound together the city and countryside.[50]

Although railroads function as physical infrastructure, transportation systems in general operate on several levels simultaneously. As systems of fixed capital, railroads were themselves socially produced through capitalist investment processes. Railroads may have appeared as disembodied objects on the landscape, but they had no existence apart from the institutions that created and governed them—railroad companies.[51] At their inception, the singular goal of railroad companies was maximizing carrying trade as a way of earning the highest return on investment. Therefore, the early investment strategies of railroad companies centered on extending their lines as far and as widely as possible. As the national rail network became complete and intra-industry competition winnowed the number of profitable lines, railroad companies became more diversified in their investments. Alfred Chandler called American railroads the nation's "first big business" because they pioneered so many of the business practices that became commonplace in corporate America.[52] Not least among these was the diversified investment symptomatic of maturing enterprises that created the multidivision corporation. Developing and operating industrial districts was a novel investment activity of the maturing railroad industry.[53] The founding of Chicago's Central Manufacturing District, for example, took place after initial investors in the Chicago Junction Railroad (and Union Stock Yards) sold out to a syndicate incorporated in New Jersey. The syndicate owned and operated "Central Manufacturing Districts" in several cities across the United States.[54] Yet even in developing industrial real estate, the interests of railroad companies in maximizing carrying trade was still evident.

Along with being physical infrastructure connecting Chicago with its expanding hinterland, and capitalist institutions bent on maximizing profit, Chicago's railroads were also industries that generated new demands for goods and services in the local economy. For example, Riley states that already in 1857 the first rolling mill was established in the city to re-roll iron rails.[55] He also enumerates thirty-five different metalworking firms in Chicago engaged in supplying railroads with freight and passenger cars. The establishment of the Pullman car works was clearly linked to Chicago's keystone position in the national rail network, as were many of the firms that set up shop in the South Branch district. Numerous railroad companies also built their own manufacturing and repair shops in the city. These and other metalworking industries, like the farm implements industry, contributed greatly to the local demand for iron and steel, which by the end of the century gave rise to Chicago's mammoth primary metals production complex.[56]

Organized industrial districts present a relatively clear relationship between railroads and the formation of Chicago's industrial landscape: railroads were the actual real estate developers. Along the suburban corridors through which the railroads were built, the relationship was somewhat less straightforward. These adjacent suburban lands typically became industrial zones developed in piecemeal fashion by scattered real estate operators. Railroad companies had a clear interest in the expansion of industries along their routes that would contribute to their carrying trade. In fact, Ebner suggests that during the early 1850s rail company management along the Chicago and Milwaukee line expressed a clear preference for connections with industrial over residential development. "Management looked askance at the prospect of transporting commuters each day into Chicago. The original plan was to concentrate exclusively on taking lucrative profits from the hauling of freight."[57] The ad hoc industrial zones that emerged along the main trunk lines surely owed their existence to the railroads, even if rail companies did not mastermind their development.

In his doctoral dissertation, Harold Mayer noted how the seven routes of railroad approach to Chicago effected the localization of industry.[58] With the exception of the Illinois Central's lakeshore route, the routes were lined with industrial sidings, punctuated by freight houses that typically served small and medium-sized industries, many of which occupied multistory loft buildings. The areas immediately adjacent to the major trunk lines, and especially near freight stations, constituted ribbon-like industrial districts, laced through the urban landscape. Clustered along these routes were industries and firms not large enough to warrant an exclusive rail siding or private railroad of their own, as did the McCormick Reaper works and large steel works—hence the dependence on common freight houses for shipments of less than a carload.[59] The industrial areas varied in width from a few blocks to a few miles, with the most extensive industrial development along rail routes paralleling the north and south branches of the Chicago River.

> The result of the seven routes has been the development of wedges or strips of industry radiating from the commercial center of Chicago. . . . The smoke and noise that in the past have been the inevitable accompaniments of railway operation, together with the advantages that the main lines offered for the location of industries, have forced residential development to take place mainly in the interstitial areas between the prongs of industry.[60]

Transportation developments are closely related to trade: improvement in the former typically stimulates increases in the latter. Not surprisingly, then, the main backward industrial linkages in nineteenth-century Chicago were with industries processing natural resource commodities traded within the city's thriving grain, lumber, and livestock markets. The flour milling industry thrived along the Chicago River waterfront during the 1840s and 1850s but declined after the Civil War, even as the city retained its position as the nation's chief grain auction market. The lumber district that sprawled for a mile along the South Branch (south of 22nd Street and west of Halsted) became home to a host of allied industries, including furniture and piano factories, wagon and shipbuilding concerns.[61] Most infamous of the processing linkages was with the livestock trade, which transformed a vast portion of the Chicago landscape in its image.

City Within the City: The Union Stock Yards and Packingtown

Chicago's livestock trade and meatpacking industry are as old as the city itself.[62] Stockyards were located around the periphery of the city along the roads used by livestock drivers. As was the custom throughout the West, tavern keepers operated most of the early stockyards, catering to transient drovers and stockmen by fencing pasture adjacent to their inns. As the railroad became the preferred method of livestock transportation during the 1850s, the advantages of having stockyards adjacent to rail lines were not lost on the railroad companies. By the mid-1860s three rail companies had constructed yards in the city, bringing the total number of operating stockyards to six by 1864.

Many voices called for consolidation of the city's livestock facilities. As urban expansion surrounded the once-peripheral stockyards, drovers and livestock brokers were forced to convey animals from market to market over rough and ungraded city streets; frequently heard were complaints of animals damaged and lost as they made the circuit. For their part, packers and wholesale livestock merchants argued that the dispersed yards made their transactions unwieldy, while the railroads complained about the expense of switching stock cars among different lines. The economic problem of a locally fragmented market was that it was, in Cronon's words, "difficult for buyers and sellers to compare the prices being offered in different yards. Financial reporters for the city's newspapers had trouble gathering information about price movements, and the inaccuracy of the resulting reports compounded the difficulties of those in the trade."[63]

Meanwhile, Chicago's rising burghers, entertaining notions of metropolitan grandeur, complained loudly about the sights and smells of hogs and cattle roaming city streets.

The Union Army's provision demands during the Civil War gave Chicago's packing industry an enormous boost, making it the largest meat-packer in the world, which only made the calls for stockyard consolidation more urgent. Fittingly, Chicago's Pork Packers' Association spearheaded an effort in 1864 that brought together the city's nine largest railroads into the Union Stock Yards and Transit Company. The consortium was led by John Sherman, whose Sherman Yards, founded in 1856 on Cottage Grove Avenue between 25th and 31st streets, were the most popular in the city.[64] He went on to become the first superintendent of the Union Stock Yards.

The company's first order of business was obtaining land to locate the stockyards away from the city: a half square mile just south of the city limits in the town of Lake. While the site is often described as being far beyond the city's reach, in fact the land had been bought and subdivided much earlier by "Long John" Wentworth, who had also operated the Brighton racetrack. In other words, the Union Stock Yards Company was not the first to see development possibilities in the seemingly valueless marshland. Far from being *terra nova*, moreover, the location of the Union Stock Yards placed them cheek by jowl with the community of Bridgeport, then home to many existing slaughterhouses along the South Branch.[65]

After five months of breakneck construction, the Union Stock Yards opened on Christmas day 1865 and were an immediate success—far beyond the dreams of investors. The original site was expanded incrementally, until, by 1900, a combination of rail yards and stock pens extended for a mile north to south and a half-mile east to west. Approximately 130 miles of railroad belt and another fifty miles of streets and alleys connected open and covered pens. For operating convenience, the yards were divided into four sections (A, B, C, and D), each assigned to specific railroads. Division B, for example, was assigned to the four eastern trunk lines entering Chicago.[66] The Yards had the capacity to handle 50,000 cattle, 200,000 hogs, and 30,000 sheep. The Union Stock Yards and Transit Company employed 1,000 people, while commission firms employed another 1,500. As all accounts suggest, the Union Stock Yards were virtually a city in themselves. They had their own police force and fire department, electric plant and waterworks, bank, hotel, amphitheater, and newspaper.[67] The Stock Yards were a marvel to Chicagoans, known affectionately as "Chicago's Pride." They also were the city's leading tourist attraction for years.

Although the Pork Packers' Association played a leading role in organizing the Union Stock Yards, initial plans did not include an adjacent industrial site for packinghouses.[68] While the Yards were under construction, however, land directly west was purchased by two land syndicates and the Union Stock Yards and Transit Company guaranteed them free access to all railroad tracks.[69] In the summer of 1868 this land was put on the market as the "Packers' Addition." The first packer to gamble his fortune on the new site was Benjamin Hutchinson, a prominent South Branch packer. Hutchinson purchased land in Packers' Addition in the fall of 1868 and proceeded to construct a large packinghouse, a smokehouse, a brick warehouse, livestock pens, and a boardinghouse for his workers. At the same time, he bought up several small firms and incorporated the Chicago Packing and Provision Company, which immediately became the largest firm in the city. Within two years three smaller plants and several slaughtering sheds catering to local butchers had located in the Addition. One observer of these developments remarked in 1870 that "without being gifted with any prophetic powers, it may be anticipated that within a few years at the utmost, the entire packing trade of this city will be carried on in the immediate vicinity of the Union Stock Yards."[70] Indeed, two years later Philip Armour relocated from Milwaukee to Packers' Addition, where he constructed a huge pork house and the first large "chill room" in the United States. Things had changed in the once "valueless marsh" southwest of Chicago. Armour paid the same price for his twenty-one-acre site—$100,000—that had been paid for the original 320-acre Union Stock Yards site in its entirety.[71] Emulators of Hutchinson's and then Armour's successes soon followed, and by the late 1870s about half of Chicago's packers, and all of the vanguard firms, had relocated to "Packer's Addition."[72] The industry as a whole experienced tremendous expansion at the end of the nineteenth century. Between 1870 and 1900 Chicago's meatpacking industry grew 900 percent. It had become the city's largest industrial employer, with thirty-nine plants and 25,345 workers, paying 10 percent of the wages and accounting for a third of the total value of Chicago's manufactured goods.

The simultaneity of Hutchinson's move from Bridgeport to Packers' Addition, his acquisition of smaller firms, and the incorporation of the Chicago Packing and Provision Company provides important clues to the changes then taking place in the meatpacking industry as a whole. During the antebellum era, in Chicago as elsewhere, meatpacking had shared with other kinds of agricultural processing a hybrid status as a commercial industry.[73] Most packers were merchants, typically in the general provisions trade, and packing was a seasonal business restricted to the winter months, when cold temperatures permitted safe handling of freshly killed meat. Wintertime packing also represented a business opportunity for merchants to employ their warehouses during the slack season. Accordingly, fixed investment in plant remained relatively small. Chicago historian Bessie Louise Pierce reported that in 1860 the total investment for the city's thirty packing firms amounted to just $155,000. During the antebellum era, meatpacking was almost exclusively confined to pork because, just as it is today, beef was largely a fresh meat product.

The 1860s were years of accelerated transformation as meatpacking made its decisive shift from commerce to industry. The transformation had several underlying causes. Like merchants and manufacturers throughout the northern United States, meatpackers shared in the general Civil War prosperity. After the war, many large packers were flush with capital for reinvestment and acquisition. Investments were channeled into new technologies for refining the "disassembly line" that was the centerpiece of the industry and for refrigeration, especially refrigerated rail transportation that would bring the dressed beef market within the packers' profit-making realm. Centralized stockyards in Chicago and the expanding rail network regularized supply conditions and permitted packers to concentrate on production and distribution. Increasingly during the 1860s, meatpacking ceased being a seasonal, sideline mercantile activity and became a full-time, year-round industrial enterprise. Accordingly, packers began devoting more attention to byproducts processing, diversifying their business as they consolidated it.

Aside from the slaughtering sheds catering to local Chicago butchers, the industry that became spatially concentrated near the Union Stock Yards in Packers' Addition represented vanguard firms incorporating (if not themselves innovating) the industry's principal transformations of the 1870s: Armour, Morris, Swift, and Cudahy. Packingtown (as Packers' Addition became known) did not merely replicate the pattern of spatially concentrated packing plants that had existed in Bridgeport before the advent of the Union Stock Yards. Packingtown became a true industrial complex or, as Jablonsky states, "a mosaic of sub districts, each created by a major packer."[74] When open land was not yet at a premium in Packingtown, packers used space extensively. Empty lots were employed as "hair fields," for example, to dry hog bristles that would later be used in paintbrushes. As industrial research continually found new uses for byproducts, each new product had its own space requirements, which often meant a separate building. By the early twentieth century, each company's subdistrict might consist of many buildings, some of them five to ten stories tall. Swift and Co., for example, occupied nearly eight square blocks between 41st and 43rd streets in Packingtown. The complex comprised thirty buildings of various sizes, including a five-story general headquarters with 2,200 employees, a print shop for producing company labels and stationery, several warehouses, and various slaughtering and processing plants. Like its major competitors, Swift developed numerous byproducts, including margarine, gelatin, furniture glue, violin strings, surgical sutures, buttons, hairbrushes, chessmen, knife handles, perfume, soap, and cleansers.

Union Stock Yards became a magnet not only for industry but also for the immigrant working-class residential community that became known as "Back of the Yards." Some of the earliest worker housing was provided by companies that pioneered the development of Packingtown. Philip Armour permitted some of his "old hands," who followed him to Chicago, to build cottages on company land at 43rd Street and Packers Avenue.[75] One block north, Benjamin Hutchinson's Packing Company erected houses for its foremen. Only skilled and supervisory workers seem to have benefited from such packer largesse. Directly east of the Union Stock Yards, moreover, in "front of the yards," the community known as Canaryville was founded as a middle- and upper-class enclave by the first generation of packinghouse owners, who wished to remain close to their factories. In the mid-1870s the Swifts, Libbys, and Hutchinsons had all settled on Emerald Avenue. The class character of the neighborhood would soon change, however. As the packinghouses prospered and the city surrounded the Stockyards district, owners and white-collar workers moved to lakefront communities further east. They were replaced in Canaryville by Irish packinghouse workers.[76]

The decisive factor in the growth of the Back of the Yards was that packers paid such low wages to their rank-and-file workers that they could not afford to live far from Packingtown. Neighborhood after neighborhood grew up in an area roughly two-and-one-half-square miles in size. European immigrants constituted the lion's share of new settlers. Irish immigrants were the first to move to the Back of the Yards in large numbers, followed by Germans. Their settlement was associated with the decline in the number of skilled workers in the industry prior to the packinghouse strike of 1886. Polish immigrants became numerous during the 1880s and 1890s, as did other Eastern European nationalities such as Lithuanians, Slovaks, Czechs, Ukrainians, and Russians. By the end of the century, Back of the Yards had become a largely Polish and Slavic community, coinciding with the expanding industry's de-skilling of its labor processes.

The neighborhoods were congested, their dirt streets lined with wooden two-flat houses. Many homes lacked indoor plumbing and few enjoyed gas or electricity. Malodorous air permeated the area. Although the Back of the Yards provided much grist for the mill of social reformers horrified by wretched living conditions, the neighborhoods were alive with institutions, formal and informal, that provided a dense matrix for social life. Chief among these was the Catholic Church. A mark of each immigrant group's arrival in Back of the Yards was the establishment of a separate parish church (and often school) to serve the specific nationality. These churches survive on the landscape as reminders of the vibrant community that once thrived in "the jungle."

The livelihoods of all revolved around the daily arrival of livestock at the Union Stock Yards. By the time the 1920 census rolled around, 75,920 people called the Back of the Yards home. A settlement school study of six selected blocks reported that 54 percent of all household heads were employed in either the Yards or Packingtown.

The simultaneity of structural and spatial change in Chicago's meat-packing industry provides a telling glimpse of the heady mix of forces that combine to produce urban space. By the time the Union Stock Yards, Packingtown, and Back of the Yards residential community achieved their zenith in the early twentieth century in what was once considered a remote, valueless marsh, urban expansion had thoroughly incorporated the area into Chicago. Although the Union Stock Yards and Packingtown were not parts of a planned industrial district, together they exemplify how "unplanned" but by no means inchoate markets for land, materials, and labor build up pieces of urban space and then literally draw the city around them. In these ways, the Stock Yards district was little different from Roseland, Grand Crossing, Brighton, and other once peripheral industrial areas of Chicago.

Model City?

In every highly organized industrial community there naturally develops a group of intermediaries or "market institutions" which function by bringing together, on the one hand, industrialists seeking suitable plant sites, and, on the other, landowners seeking to sell their holdings. Thus by providing facilities for exchange, industry is diverted to certain favored locations and the general character of the surrounding community is consequently influenced.

—W. N. Mitchell and M. J. Jucius, "Industrial Districts of the Chicago Region and Their Influence on Plant Location," *Journal of Business* 6 (1933): 139.

The accumulated wisdom of urban land-use theory has cities growing outward from an initial core area to an ever-retreating suburban periphery, with the gradual filling of interstices and eventual emergence of a multi-nucleated metropolis. But developments in Packingtown, Calumet, along the South Branch, and in the planned industrial districts suggests that the formation of Chicago's urban landscape was not a singular spatial process but was episodic and multinucleated from the outset. Firms in emerging industries and old firms in existing industries typically located new factory investments outside the city's built-up area. These location decisions were guided by a host of land developers, especially in outlying suburban areas, who acquired parcels near existing transportation routes and, in order to realize a return, competed to direct new economic growth to their properties. The logic of this competition among land developers helps account for the multinucleated pattern of urban spatial development, because the best land deals, then as now, can typically be made at some distance from thriving and established centers.

Much that seems true about the role of industrial real estate developers in shaping the urban landscape has been widely documented for developers of residential areas, particularly suburban communities. Whereas residential developers might tout their proximity to parks and churches, the field of industrial dreams offered its own set of amenities in infrastructure, services, and distance from city taxes and nuisance ordinances.[77] A host of mediating actors and institutions came to organize the housing market, to the point where real estate agents and mortgage brokers enjoy a nearly unassailable place in residential property transactions.[78] Similarly, the industrial landscape was not shaped by industrialists making bold, solitary location decisions. That the history of exclusively residential suburban development has received vastly more attention no doubt helps account for the rather poorly understood role of land developers in suburban industrial areas of cities during the nineteenth century.

The multinucleation evident in Chicago from the middle of the nineteenth century became camouflaged to later observers, including the Chicago School of Sociology, because the city enveloped once outlying settlements. The ahistorical nature of their social research kept them from seeing the more varied story within the urban landscape. It is somewhat ironic that the concentric ring model of orderly outward development became so closely identified with Chicago, because, as Mayer and Wade point out, "the centrifugal movement in Chicago began with the first generation of urban dwellers."[79] Also puzzling is that the Chicago School of Sociology would promulgate the notion that Chicago's land-use patterns were the cumulative outcome of individualistic competition, given the prominence of mediating market institutions in the city's economy, not the least of which was the colossal Board of Trade! Their market triumphalism was heavily buttressed by the kindred Chicago School of Economics that ascended to prominence at roughly the same time.

Chicago may not have been "one of the most planned cities of the modern era," as Suttles put it, but neither was its urban industrial landscape a proving ground of atomistic social and economic dynamics.[80] Leaving their marks on the landscape was a constellation of political and economic actors that included far-sighted real estate developers, transportation investors, and, eventually, professional planners. In these ways, perhaps Chicago was, if not a paradigm of urban development, then at least a thoroughly typical ideal type.

5

A City Transformed
*Manufacturing Districts and Suburban
Growth in Montreal, 1850–1929*

ROBERT LEWIS

etween 1850 and 1929 the economic and social geography of
Montreal, like that of most other North American cities, was
transformed. From a small city of 57,000 bordering the St.
Lawrence River in 1851, it had grown to a large, multinodal
industrial complex containing almost 1 million inhabitants by
the time of the Great Depression. It is generally accepted that
the realignment of the social geography of cities accompanied
their increasing scale. Building on the ecological models
inspired by Burgess and McKenzie, research into the development of urban space centered on the class and ethnic rearrangement of people and homes. The classic statements stressed the
growth of middle-class residential districts on the expanding
suburban fringe and the concentration of the working class and
immigrants in the central city.[1] Workers needed to be close to
their place of work and the middle-class independent suburbs
needed to be connected to the downtown area by transportation filaments—the horse-drawn trolley at the middle of the

Reprinted with minor revisions from Robert Lewis, "A City Transformed: Man-
ufacturing Districts and Suburban Growth in Montreal, 1850–1929," *Journal of
Historical Geography* 27 (2001): 20–35, by permission of the publisher Academic
Press.

nineteenth century, the electric streetcar after 1890, and the auto by the 1910s. But do these generalizations hold true? Several authors argue that the social geography of the North American city does not fit the conventional picture and point to other important themes: the importance of working-class and immigrant suburbs in the North American urban landscape after 1850, the continued existence of central-city bourgeois neighborhoods, and the highly differentiated and fine-grained character of the class, occupational, and ethnic residential landscape.[2]

This rise of the industrial city was also linked with the reordering of the urban industrial landscape. As noted in Chapter 2, the received view holds that production was concentrated in the city center before World War I and only began to move to suburban areas and satellite towns of North American metropolises in the 1920s.[3] Although studies show that different economic functions in the central city separated into distinct districts after 1850, the prevailing view of the internal urban industrial geography emphasizes the monocentric character of industry. Production remained in the core before the 1920s, it is generally argued, because several obstacles to suburbanization—the need to reduce transportation costs, the location of labor and transportation facilities, and a small market—forced firms to depend on agglomeration economies and therefore cluster in a compact industrial node in the city core.

Evidence from studies of several cities, however, some of them included in this book, suggests that distinct noncentral industry existed at a much earlier date.[4] Indeed, between 1850 and 1929 the tightly bound commercial city constantly uncoiled. Each wave of economic growth was associated with the reshaping of the urban industrial fabric. Even though centrally located industrial districts grew at an impressive rate, another fundamental trend, the emergence of industrial districts on the moving urban frontier, affected the city's geography after 1850.[5] This process of industrial suburbanization shaped the geography of Montreal between 1850 and 1929. Three fundamental forces were at work: different production pathways presented the opportunities for firms to decentralize their production functions; the workings of the property market opened up suburban land for development; and growth politics created the ideological and material foundations for suburbanization. The history of Montreal's industrial suburbanization is examined here through two waves of development. These are reconstructed through an analysis of suburban manufacturing districts in 1861, 1890, and 1929, based on data drawn from Montreal's water tax assessments (rôle d'évaluation).[6]

Capitalist Industrial Suburbanization

Capitalist industrialization in Montreal between 1850 and 1929 transformed a small commercial city, hugging the shores of the St. Lawrence River, into a large multinodal industrial metropolis. In each wave of growth the locational coordinates of industry were reformulated. As in other American and Canadian cities, the dynamics of technological change and the redrawing of the division of labor led to the search for new manufacturing spaces that opened up the urban fringe and the suburbs for development. From the 1850s, manufacturing firms often led the expanding urban frontier out from the built-up city. For an increasing number of firms, a suburban site was part of a logical spatial strategy involving better labor relations, lower production costs, and the ability to construct a new work regime. Accordingly, greenfield sites in rural areas and existing "satellite" towns became part of manufacturing districts extending out the built-up city. In nearly all cases suburban manufacturing located along transportation corridors radiated out from or encircled the city center. Although Montreal's urban growth differed from that of other North American cities because of the specifics of its industrial structure, land development policies, different ethnic groupings, and political alliances, it was propelled, as with other cities, by the movement of industry from the city core to the fringe and industrial suburbs.

Three sets of dynamics shaped the city's industrial geography. First, a set of firms employing different pathways of industrial growth and seeking a new set of locations spurred industrial suburbanization. New fixed capital formation, the expansion and remodeling of existing plants and infrastructures, and the introduction of new technologies, work methods, and propulsive industries, a substantial share of which were channeled into different parts of the city, characterized each wave of investment. Associated with the long-term, but lumpy, growth of industrial investment was the development of a varied set of production trajectories. Despite the growing scale of industry, the drive to industrial concentration, and the widespread use of high-volume, repetitive manufacturing methods, the actual pathways opened to manufacturing firms were extremely diverse and involved the deployment of varied technological and organizational forms, distinct strategies of marketing and labor control, and a range of business-political relations.[7] The different trajectories of growth and development created a range of social and technical relations of production that in turn produced new sets of place-specific work environments.

Second, capitalist land development practices opened peripheral land for industrial and residential use. In Montreal this can be traced through the actions of a series of actors. Land developers, businesses, religious and business organizations, local municipalities, and transit companies opened land for industry throughout the city fringe and the suburban districts. Industrial demand for large lots, new spaces to install new production methods, access to railroads and waterways, and escape from the diseconomies of central sites (congestion, labor unrest, and high property values) encouraged the property sector to assemble peripheral industrial land that fanned out from the city core and extended along the major transportation routes. Typically, suburban industrial nodes established in one building cycle generated the basis for subsequent extension along the expanding industrial frontier in the following cycle. Sometimes, however, industrial districts leaped into uncharted territory, when the property industry was able to package and shape the appropriate forms of land.[8]

Third, the creation of suburban industrial locations could not have taken place without the development of specific forms of growth politics, involving alliances between local economic groups and various levels of the state. These alliances continually shifted and were at odds with one another; nonetheless, all sought the same ends and were responsible for laying down the physical, ideological, and legal structures necessary for industrial growth and the rearrangement of the urban fabric.[9] The making of suburban industrial spaces in Montreal by the local actors took many forms. In the first place, coalitions consisting of property promoters, local politicians, and businesspeople developed new areas for industrial settlement and for associated working-class residences. Second, local alliances built new and remodeled existing infrastructures to reproduce industrial class relations in space through the defense of existing capital embedded in the urban landscape and the generation of new urban resources through which they could privately profit, and in the process laid the basis for the emergence of new spaces on the city fringe amenable to industrial development. Finally, despite disagreements over the particular form that capitalist urbanization should take, local elites, with support at the provincial and national levels, shaped the legal and material foundations of suburban growth through control over the ideology of local urban expansion. New spaces on the city fringe amenable to industrial suburbanization emerged as one result of these processes.

Early Beginnings of Montreal's Industrial Suburbs

Montreal's multinodal industrial complex formed as distinct and specialized suburban industrial districts at different points in time between the mid-nineteenth century and the onset of the Great Depression. Its origins can be identified with the burst of growth beginning in the late 1840s and continuing through the end of the century, and the expansion occurring between the 1890s and 1929 elaborated a more fully developed metropolitan structure. In each phase, the urban frontier moved farther out, older industrial suburbs were incorporated into the built-up area, and new ones formed on the city's fringe.

The wave of capitalist industrialization occurring after the late 1840s initiated the suburban movement of Montreal's industry. Old Montreal, the original core of the city, contained most of the city's economic activity before mid-century. Over the course of the century, it experienced great changes, notably the concentration of small, labor-intensive firms in the clothing, shoe, printing, and cigar industries, and decline relative to other city locations.[10] Firms moved into adjacent areas, producing a surrounding belt of inner-city industrial districts (St. Antoine, St. Lawrence, and St. Jacques) (see Figure 5.1). These districts specialized in clothing, printing, and other light industries and took advantage of agglomeration economies derived from Old Montreal. This extensive, growing, and centrally located industrial complex was only one feature of the city's industrial geography, however. Suburban industrial districts that emerged in the western and eastern fringes of the city were equally important. By the

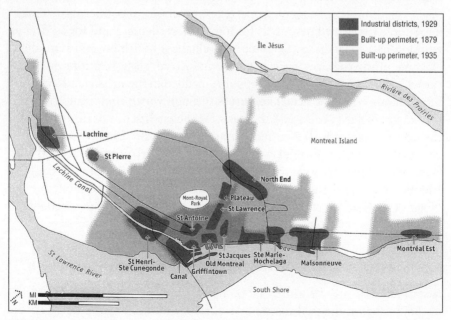

FIGURE 5.1. Montreal's Manufacturing Districts, 1929.

TABLE 5.1. Suburban Manufacturing Districts in Montreal, 1861 and 1890

Manufacturing district	Date of emergence	No. of firms		% share of city rent		Median rent ($)	
		1861	1890	1861	1890	1861	1890
West End							
Canal	1840s	40	68	26	12	550	310
St. Henri	1870s		88		11		150
Lachine	1880s		13		2		60
East End							
Ste. Marie	1850s	40	127	6	12	44	120
Maisonneuve	1880s		10		3		90
Suburban share (%)		13	23	32	40		
City total		631	1,352	100	100	100	200

Source: Rôle d'évaluation for Montreal and surrounding towns, 1861 and 1890.

1890s a more elaborate industrial geography was evident, as new districts emerged and existing ones expanded.

After 1850 capitalist industrialization produced new industries and new competitive pressures.[11] An array of industries spanned the gamut from large-scale, mechanized firms specializing in high-volume production to the small artisan workshop and the sweatshop. This complex industrial structure was manifested in considerable differences in capital outlay, scale, labor process, markets, and mechanization, thus producing wide production-cost variations and workspace requirements. In the face of these processes, one option open to some firms was to seek out greenfield sites in fringe and suburban districts, both to reduce costs and to implement competitive production techniques in new industrial spaces. The result was the revaluing of location within the city, which led to the emergence of fringe and suburban industrial districts.

The Canal district was the key fringe industrial cluster emerging after mid-century (see Table 5.1 and Figure 5.1). It was the locus of a technologically advanced, hydraulically based, energy-intensive form of production, and featured growth in the number of firms of all sizes, changes to the industrial and technological organization of work, and firms from the most important propulsive sectors of the day, food and metal (see Table 5.2).[12] These large firms looked for some combination of cheap land and large lots, an escape from the labor problems of the center, access to better transportation facilities, and greenfield sites on which to install new forms of machinery and production rationalization. By the 1860s the Canal

TABLE 5.2. Industrial Structure of Selected Suburban Manufacturing Districts

1861

Industrial district	Firm median rent ($)	Rent distribution (%)			Three largest industries (by $ rent)	
		Large >$299	Medium $50–299	Small <$50	Industries	Total %
Canal	550	72	10	18	Locomotive, flour, iron and steel	63
Ste. Marie	44	8	32	60	Beverage, rubber, baking	55
City total	100	23	48	29	Clothing, shoe, beverage	29

1890

Industrial district	Firm median rent ($)	Rent distribution (%)			Three largest industries (by $ rent)	
		Large >$449	Medium $100–449	Small <$100	Industries	Total %
Canal	310	46	23	31	Locomotive, metal goods, flour	51
St. Henri	150	29	44	27	Cotton, iron and steel, metal goods	49
Ste. Marie	120	21	39	40	Locomotive, rubber, tobacco	47
City total	100	24	51	25	Clothing, shoe, beverage	29

1929

Industrial district	Firm median rent ($)	Rent distribution (%)			Three largest industries (by $ rent)	
		Large >$1,999	Medium $400–1,999	Small <$400	Industries	Total %
St. Henri	1,400	38	34	28	Cotton, metal goods, iron and steel	44
Lachine	1,800	42	28	30	Machinery, metal goods, locomotive	78
Montréal Est	11,800	94		6	Cement, locomotive, petroleum	90
North End	700	19	45	36	Clothing, paper, chemicals	41
City total	800	26	48	26	Clothing, metal goods, locomotives	25

Source: Rôle d'évaluation for Montreal and surrounding towns, 1861, 1890, and 1929.

district and the adjacent district of Griffintown, which was older and more centrally located, comprised a large industrial complex of diverse yet linked firms in the West End. Although the relative importance of the Canal district diminished in the last quarter of the century, the size and scope of the complex continued to grow. The emergence by the 1880s of industrial suburbs (St. Henri and Ste. Cunégonde) and a satellite town (Lachine) farther west heralded the continuation of existing industrial sectors (food and metal) and the establishment of new ones (textiles and chemicals).[13]

A fledgling industrial district also developed in the East End. Although it was small and centered on a small range of industries at mid-century, rapid growth occurred over the following decades: new propulsive industries appeared (cotton, tobacco, and oil cloth), and Ste. Marie, the original core, and Hochelaga, a new industrial suburb, emerged (see Tables 5.1 and 5.2 and Figure 5.1). Having a more bipolar industrial structure than the West End, Ste. Marie and Hochelaga were characterized by the coexistence of several large, high-volume, mechanized firms, such as Molson Brewery and Canadian Rubber, which expanded and reorganized production during the period, and of many proprietary small-scale firms that employed few workers, served local markets, and operated with simple production methods. By 1890 the East End districts formed a distinct industrial nucleus accounting for 15 percent of the city's total rent.[14]

The development and success of these districts was tied to the construction of suburban working-class housing. The number of working-class households in the outlying districts of St. Ann and Ste. Marie wards, for example, greatly increased after 1848, while new industrial suburbs began to dot the metropolitan landscape by the 1870s (see Table 5.3).[15] These new fringe residential districts coalesced around industrial nodes, and their occupational structure reflected the employment demands of the districts. The metalworking and locomotive districts of the West End, for example, housed high concentrations of molders, machinists, engineers, baggage men, and fitters. Occupational specialization was paralleled by ethnic segregation; extensive French settlements existed in the east and English ones in the west, although segregation could occur within districts at the street level.[16] The territorial clustering of suburban factories, working-class residential settlement, and occupational and ethnic specialization was replicated across the expanding industrial fringe.

Montreal's property market tied together the industrial and residential functions of the peripheral districts. Shortages of cheap housing were a continual problem for Montreal's workers, even after the pressure on inner-city housing was relieved after 1850, as wealthy Anglophones moved

TABLE 5.3. Social Indicators of Newly Developed City Fringe Districts and Suburbs, 1848–1881

	No. of households			French households (%)		Mean residential rent ($)		
	1848	1861	1881	1848	1861	1848	1861	1881
City fringe districts								
Western St. Ann	266	1,420	3,415	29	22	51	45	50
Eastern Ste. Marie	204	654	3,548	50	93	37	33	39
Upper St. Louis and St. Jacques		95	1,673		60		53	68
Upper St. Antoine and St. Lawrence	275	772	2,184	22	19	182	232	324
Industrial suburbs								
Hochelaga			836					64
St. Gabriel			1,515					60
St. Henri and Ste. Cunégonde			2,368					48
Northern suburbs			1,777					57
City fringe total	745	2,941	10,820	32	38	120	91	105
Suburban total			6,678					58
City total	5,320	12,330	33,350	42	46	79	71	84

Source: D. Hanna, "Partage social et partage de l'espace à Montréal, 1847 à 1901" (Rapport d'etape au Fonds F.C.A.R. Québec, 1986), and S. Olson, "Partage social et partage de l'espace à Montréal, 1847 à 1901" (Rapport d'etape au Fonds F.C.A.R., Québec, 1986).

to the new bourgeois district of upper St. Antoine, leaving their houses to be subdivided into tenements. But a growing population, coupled with devastating fires, the encroachment of the expanding central business district, low returns on centrally located working-class housing, low wages, and the move of industrial employment to the fringe forced many to flee the expensive inner city and seek suburban housing.[17] In contrast to large contractors and building companies that built housing for the middle class, small contractors and speculative developers constructed a suburban quilt of cheap housing. Usually they followed the exodus of workers to the city fringe, suburbs, and satellite towns, though at other times they established a working-class community that attracted firms wishing to locate on the periphery. By the 1850s working-class housing and industrial sites were constructed in the Canal district by three groups: the state redeveloped the Lachine Canal for industry; the St. Sulphician seminary set out extensive tracts of land for housing and industry; and a merchant-manufacturing clique subleased the canal lots from the government, bought up and sub-divided surrounding land, and established manufacturing plants along the canal.[18] Differing in the details, the same basic process of the construction of cheap, two-story row houses on small lots close to firms was duplicated throughout the fringes of Ste. Marie and the industrial suburbs of St. Henri, Ste. Cunégonde, and Hochelaga. Although the mismatch between population growth and house building remained unsolved throughout the period, the growing number of small, cheap dwellings on the periphery made the suburbs an attractive area for industry and workers.[19]

Local growth alliances were active in the creation of the suburban indus-trial landscape. Through a wide array of infrastructural improvements, land development practices, and booster strategies, Montreal's political and economic elites were responsible for the construction of peripheral land amenable to industrial development.[20] The establishment of Mon-treal's Harbour Commission in 1830 allowed for the organization, financ-ing, and construction of new port facilities stretching out from the city core to what were to become new suburban districts.[21] Although the Harbour Commissioners had more immediate commercial objectives in mind, an unintended result was the creation of a physical structure that supported the movement of the city's industry to new suburban locations after mid-century.[22] City support for the establishment of an urban railway network also provided important facilities for manufacturers throughout the city. The development of an extensive railway system along the Harbour Com-missioners' land, the crisscrossing of the city by the railway tracks of the Grand Trunk and Canadian Pacific railways, and the building of large freight yards to the west, east, and north allowed manufacturers to move their plants out to noncentral locations and gain excellent access to national and international markets through a spur line or a location near the freight yards. Moreover, the Grand Trunk fabricating and repair shops at Point St. Charles after 1856 acted as a nucleus for the formation of an extensive set of industrial linkages and a large working-class residential district in the Canal district.[23]

Industrial Suburbanization and a Multinodal Metropolitan Structure

By 1929 a more elaborate industrial geography was evident. Larger lumps of capital, the arrival of new propulsive industries, a more active local state, the immigration of people from Eastern and Southern Europe, a more extensive and corporate-oriented land development process, and the laying

TABLE 5.4. Suburban Manufacturing Districts in Montreal, 1890 and 1929

Manufacturing district	Date of emergence	No. of firms		% share of city rent		Median rent ($)	
		1890	1929	1890	1929	1890	1929
West End							
Canal	1840s	68	59	12	8	310	2,400
St. Henri	1870s	88	140	11	12	150	1,400
Lachine	1880s	13	43	2	9	60	1,800
East End							
Ste. Marie	1850s	127	145	12	7	120	780
Maisonneuve	1880s	10	105	3	6	90	1,000
Montréal Est	1900s		18		9		11,800
North End	1900s		292		8		700
Suburban share (%)		23	32	40	59		
City total		1,352	2,474	100	100	200	800

Source: Rôle d'évaluation for Montreal and surrounding towns, 1890 and 1929.

down of a new built environment shaped the nature, direction, and rate of industrial suburbanization. In step with the uncoiling of residential districts, new infrastructures, and transportation networks from the built-up core, industry continued to seek suburban locations, creating a multinodal metropolitan geography. The suburban manufacturing districts created in the earlier growth phase had been enveloped by the expanding city and new ones formed on the periphery (see Table 5.4 and Figure 5.1). For the most part, the early suburban industrial districts maintained the industrial mix: Canal was dominated by a few large food and metal firms linked to wider international markets, while Ste. Marie extended its line of large consumer firms. The industrial districts created at the end of the last round of investment expanded; St. Henri, for example, had 140 firms, including some of the largest in the city, and 12 percent of the city's rent in 1929.

The peripheral manufacturing nodes emerging on the suburban fringe and containing propulsive industries such as steel, oil refining, and chemicals were organized along different lines from each other and from earlier industries (Table 5.2). To the west, the development of satellite towns such as Lachine and Ville St. Pierre centered on the implantation of large transportation and steel plants; they were typically nonlocally owned (some by Americans, others by corporate capital centered in Montreal and Toronto) and employed the latest technologies and work methods.[24] In the east, the earlier manufacturing complex expanded as Maisonneuve (metal, food, and shoes) and Montréal Est (cement and oil refining) grew. After 1900 a new manufacturing cluster emerged in the northern part of the city alongside the Canadian Pacific Railway lines from Ste. Marie to north of Outremont; this district housed numerous small firms in the paper, food, chemical, and clothing industries.[25] The reorganization of Montreal's industrial geography, however, did not produce a simple polarization of scale and industrial type between core and suburb. Large firms that employed the latest manufacturing methods were usually the basis of industrial suburban development, but a range of firms catering to highly differentiated markets deploying different production methods, and with a wide range of investment capabilities, also sought suburban locations to take advantage of the agglomerative economies initiated by the large firms in propulsive industries.

The emergence of a new political and ideological climate made possible industrial suburbanization and the external economies that it generated. Until the 1880s, the small elite that dominated municipal politics concentrated on the provision of patronage and resources for the Anglophone bourgeoisie. With urban expansion and suburban annexation, the ethnic and class balance changed. The ascendancy of French Canadians to municipal power and their search for individual profits through city institutions and urban growth resulted in the creation of new local institutions and alliances, battles organized along the lines of reformism and "bossism," and a more active intervention in the administration of metropolitan growth. The new municipal leaders "took pride in Montreal's growth and never entertained any arcadian ideas that growth should be tempered." To this end they advocated suburban residential expansion, the remodeling of infrastructures, limited regulation of local utilities, and additional annexation.[26] Although they were limited in their achievements, the support of relatively unbridled urban liberalism and regulated suburban expansion provided the political context for more industrial and working-class suburbanization.

As in the earlier bouts of industrial suburbanization, the success of an industrial district depended on its functional integration into the orbit of the metropolis and access by workers to suburban employment opportunities. The ability of the building industry to construct adequate housing was vital for the continued growth of a substantial suburban labor pool. In the first decades of the twentieth century, existing residential areas such as Lachine and St. Henri continued to attract industry and workers, while new working-class districts followed industry out to greenfield areas along major transportation networks; working-class suburbs such as Verdun in

TABLE 5.5. Population of Selected Suburban Working-Class Districts, 1881–1931

Districts	1881	1901	1931
West End			
St. Henri	6,415	21,192	30,094
Ste. Cunégonde	4,849	10,912	19,249
St. Gabriel	4,506	9,986	19,873
Lachine	2,406	5,561	18,630
Verdun		1,898	60,745
East End			
Hochelaga	4,111	12,914	21,838
Mercier/Montréal Est	1,114	2,519	22,639
Maisonneuve		3,958	30,165
North End			
St. Jean Baptiste	5,874	26,754	27,379
St. Louis	751	10,933	21,827
Villeray[a]		3,960	71,821
Delormier		1,279	30,960
Rosemont		315	43,964

Source: Canada, *Census of Canada, 1941* (Ottawa, 1944); Y. Lamond, *La culture ouvrière à Montréal, 1880–1920* (Quebec City, 1982), tables 2 and 3; G. Lauzon and L. Ruellard, *1875/Saint-Henri* (Montreal 1985), 11.
[a]Includes Bordeaux, Cartierville, and Ahuntsic.

the west, Rosemont in the north, and Mercier and Montréal Est in the east were important features of the urban fringe (see Table 5.5). Although many jobs for women and children remained centrally located, increasing employment for women in suburban rubber, shoe, tobacco, and textile firms was an impetus to working-class suburbanization.[27] While the building industry continued to construct cheap working-class housing close to suburban factories, suburban working-class districts also developed in response to the lengthening of the journey to work. The trolley permitted greater work-home mobility because its greater speed and passenger capacity widened the area that the intra-urban transportation system could reach, and it allowed a better integration of suburban and rural areas into the central city.[28] As industry migrated to the fringe, nevertheless, workers sought homes close to their suburban place of work because few of them could afford the trolley.[29]

Montreal's ethnic division of labor influenced the social geography of the suburbs. Although an important east-west divide existed, with French speakers on the east and English speakers on the west, this pattern was not absolute. Large French-speaking communities occupied the western industrial suburbs of St. Henri, Ste. Cunégonde, and Lachine, while clusters of people of British and Irish origin lived in the northeastern suburbs. The development of other ethnic settlements throughout the city complicated the simple division of the "two solitudes" (the Anglophone west and the Francophone east). Jews, who formed the largest group outside of the British and the French, concentrated heavily in the garment corridor leading out of the central business district. By World War I, however, a significant share of the Jewish population had settled in the Mile End manufacturing district in the city's north end, close to the decentralizing garment firms and other employment opportunities. This alliance between residential suburbanization and manufacturing employment was also true for other groups. In the last decade of the nineteenth century, Italian settlements had developed in the city's West and East Ends, and by World War I an extensive cluster emerged in the North End to take advantage of "free cultivable land in the city outskirts" and to be close to the Angus car shops, the city's largest employer of Italian labor.[30] Similarly, Ukrainians, most of whom worked in the metal sector, had large settlements close to the two large railway shops in the southwest and north of the city.[31]

An assortment of large development companies, small builders, street railway companies, and suburban municipalities transformed vast tracts of rural land to urban uses and contributed to the formation of this suburban mosaic.[32] The growth of the city thus produced a more fragmented social space, as new class formations appeared, the ethnic balance changed, and the spatial limits of the housing market grew. As the property industry created a mosaic of class and ethnic suburban neighborhoods, it became instrumental in the growth of industrial districts on the urban fringe.

Maisonneuve is the best-documented example of the alliance between industry, developers, and politicians in making an industrial suburb. From the 1890s the systematic application of an aggressive industrial policy and the creation of "model" working-class housing by a land/manufacturing clique who controlled the town council were instrumental in the growth of Montreal's suburban "Pittsburgh of Canada." The close cooperation of elites involved a policy of municipal subsidies, the installation of vital infrastructures, an extensive advertising scheme, and working-class housing. As a result, Maisonneuve grew from a sparsely settled part of the city fringe in 1890 to one of the largest industrial complexes in Canada by World War I.[33]

Another major round of public and private investment after 1900 made possible the kind of suburban developments that took root in Maisonneuve. Improvements continued to the harbor and the railways, establishing another basis for industrial suburban growth in the West and East Ends. Large-scale improvements to the harbor in Maisonneuve and Montréal Est attracted large firms such as St. Lawrence Sugar and Canadian Vickers to the eastern suburbs, while the remodeling of the Lachine Canal area provided western suburban firms with more convenient access to the central harbor.[34] New railway facilities encouraged industrial suburbanization; railway lines that led out from the central harbor, connected with the eastern marshaling yards and the Angus locomotive shops, fanned out to the northwest and skirted the built-up area of the city, drew firms to the expanding industrial districts of Ste. Marie, Hochelaga, Montréal Est, and Maisonneuve, and to the North End after 1900.[35] The introduction of electric power in the 1880s extended the industrial opportunities of the suburbs, because it gave rise to new propulsive industries, permitted the internal reorganization of factories, and allowed for the expansion of, and greater mobility within, the built-up area of the city.[36] The channeling of large public and private investments into an array of heavy-duty infrastructures provided indirect generative potential for the creation of suburban industrial districts and created new-built-environment assets throughout the city that firms treated as economies of agglomeration.

Far from being a monocentric city, nineteenth- and early twentieth-century Montreal was a large manufacturing complex composed of a set of linked yet specialized industrial districts bound by networks and by flows of capital that local alliances managed. The formation of industrial districts in the west (Canal, St. Henri, and Lachine), the east (Ste. Marie, Maisonneuve, and Montréal Est), and the north are evidence of substantial industrial suburbanization between 1850 and 1929. The Montreal case, along with the creation of fringe manufacturing complexes in other Canadian and American metropolitan areas, highlight the importance of industrial suburbanization for nineteenth- and early twentieth-century urban development. Multiple nodes of industrial development characterized the city; this highly evolved spatial division of labor was not predicated upon a dichotomous world of small/large firms and core/periphery locations. At all levels of industrial activity—the factory, the district, and the metropolis—there was constant pressure to restructure the nature of work and its associated power relations. The development of metropolitan Montreal rested on several critical features. First, productivity increases due to technological change, a widening division of labor, more efficient distribution networks, and expanding demand provided opportunities for firms to grow and move. Second, the development of different productive strategies allowed firms and industries to capture and take advantage of the wider production and urban system. Finally, the changing character of the industrial system forced manufacturers to seek out new industrial spaces.

A distinguishing feature of Montreal between 1850 and 1929 was the malleability of its industrial and social spaces. These spaces were differentiated as part of a division of manufacturing in urban space as a whole, which was made possible only because parts of Montreal's space could be molded and reconstituted in different forms at different times. Consequently, urban industrial districts varied in terms of the number and density of establishments, the mix of industrial sectors, their dominant production formats, and their relation to labor and housing markets. Yet urban space also had a rigidity that forced a sclerosis of capital and industrial activity. Each new burst of economic growth—with investment in new growth industries, the laying down of new infrastructures, demands for a different labor force, and the opening and development of suburban land— installed a new industrial geography on the urban landscape that did not eliminate existing industrial districts. Capital, especially large chunks of it, froze in the urban landscape, and formed the base upon which new flows of capital would agglomerate. As a result, the locational possibilities open to manufacturers produced a series of industrial complexes in the nineteenth and twentieth centuries. Suburban manufacturing districts as diverse as the Canal, Maisonneuve, and the North End were generated by powerful economic dynamics that blossomed into a range of production formats, regulated by cyclical flows of industrial and fixed-built-environment investment, supported by the creation of highly segmented and structured social spaces, and managed by local alliances that orchestrated and mediated the urban political and social context.

6

Industry Builds Out the City

The Suburbanization of Manufacturing in the San Francisco Bay Area, 1850–1940

RICHARD WALKER

The San Francisco Bay Area provides a clear example of industrial dispersal creating the sprawling form of the American metropolis. Manufacturing began its outward march from the earliest days of the city's industrialization, establishing major nodes of activity south of Market Street in San Francisco, then across Mission Bay after the Civil War, moving beyond the city limits to South San Francisco by the turn of the century, then over to Oakland and the Contra Costa shore on the east side of the bay. Along with the dispersal of industry came a steady expansion of working-class residential districts, tied to manufacturing and warehousing districts by time, income, and transit limitations on employees. All this happened long before the explosive growth of Silicon Valley fifty miles south of San Francisco after World War II. Contrary to the conventional view, there was not a single industrial core in the Bay Area that decentralized after World War II, and industrial development was just as important as residential suburbanization (if not more so) in driving the outward flow of urban growth.[1]

Reprinted with minor revisions from Richard Walker, "Industry Builds Out the City: The Suburbanization of Manufacturing in the San Francisco Bay Area, 1850–1940," *Journal of Historical Geography* 27 (2001): 36–57, by permission of the publisher Academic Press. The section on San Francisco has been completely revised.

The primary cause of industrial decentralization has been industrialization itself, chiefly the sheer growth of manufacturing and distribution and the search for generous spaces in which to build factories and warehouses. This is an expansion that has been punctuated by change in the nature of industry, marked by episodic outbreaks of new sectors in new places. The main turning points in our narrative are the burst of industrialization after the Civil War and the reconfiguration of industry around the turn of the nineteenth century; the 1880s and World War I are also significant moments. Cheap land has always been a draw for building at the urban fringe, as has being surrounded by similar and complementary industry—often forming distinctive industrial districts within the metropolis. If land can be packaged in an industrial park, or in a political jurisdiction dominated by industrialists, all the better. Better still if labor and unions might be contained as well outside their urban strongholds. Nature has played a minor part in driving industry outward, particularly the great earthquake of 1906 and the wide expanse of San Francisco Bay, and transportation improvements have been significant, if secondary, in facilitating the outward spiral of the city.[2]

San Francisco: Gold Rush to Industrial Hub

The Gold Rush of 1848–55 propelled San Francisco into the first rank of urban places in the United States, with a population of 50,000 in 1855 tripling to 150,000 by 1870. In addition to being the hub of the California mining region, San Francisco served as the principal mercantile and financial center for the transmontane West. For half a century, it had no major rivals closer than Chicago. The city's merchants, bankers, bonanza kings, and railroad titans became lords of all they surveyed, reigning over an empire stretching from Alaska to Mexico.[3] In the Gold Rush era most industrial activity sprang up in outlying parts of the Bay Area, which was knitted together by a network of ferries and produce boats. Most of this was resource extraction and processing, such as quicksilver (mercury) mines, smelters, logging, and lumber milling, and agriculture and food processing, such as wheat fields and flour mills, cattle ranches and slaughterhouses, vineyards and wineries.[4] San Francisco itself had only a small amount of manufacturing around Yerba Buena cove, the heart of the Gold Rush city. The percentage of total regional manufacturing lying outside San Francisco in the 1850s would not be reached again until World War I (see Figure 6.1).

According to the dominant view, San Francisco was a commercial rather than an industrial city, because the percentage of workers employed in

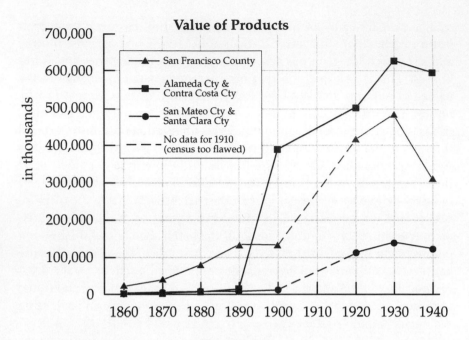

Value of Products

- ▲ San Francisco County
- ■ Alameda Cty & Contra Costa Cty
- ● San Mateo Cty & Santa Clara Cty
- No data for 1910 (census too flawed)

in thousands

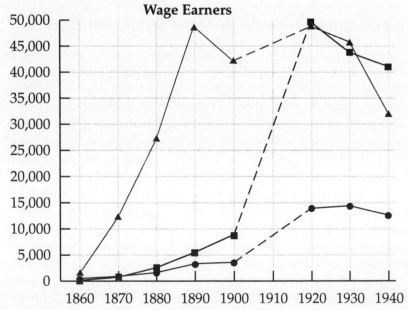

Wage Earners

FIGURE 6.1. Bay Area Manufacturing, 1860–1940.
Source: U.S. Census of Manufacturers.

manufacturing was lower than that in eastern cities of its size. Manufacturing, it is argued, was limited by the scale of demand, scarcity of labor, and lack of fuel, and was too dependent on natural resources and the local market.[5] Percentages are misleading, however, because San Francisco's mercantile empire was so much larger than any other comparable city's. Jobs in trade and transport were so plentiful that industry may appear relatively less impressive, but industrial activity began to quicken in the Civil War years and grew tremendously after that, whether measured in output, employment, or number of firms (Figure 6.1). Industrialization occurred in two great economic upswings, 1862–75 and 1878–93, with severe downturns in the depressions of 1875–77 and 1893–95, which hit the city hard. Things had revived smartly between 1896 and 1906, when the great earthquake hit. San Francisco's manufacturing workforce and output doubled in the 1870s, peaked as a proportion of all workers in 1880, and ranked ninth in the United States that year. By 1880 manufacturing occupied a third of San Francisco's workforce, counted for two-thirds of statewide employment and value-added, and exceeded the output of all other western cities combined.[6] This brought about a striking regional *centralization* of industry in San Francisco, as the city's manufacturers sold their goods around northern California, the West, and even abroad. An aerial view of the city from the early 1880s already shows smoking factories ringing the city (see Figure 6.2).[7]

Despite some disadvantages, such as high fuel costs, San Francisco's industrialization benefited from favorable supplies of capital and labor. On the one hand, industry was force-fed by locally accumulated capital. Every big investor—Billy Ralston and Charles Crocker, for example—had his fingers in a dozen pies. On the other hand, scarcity of labor was offset by rapid in-migration, the end of placer mining, and completion of the transcontinental railroad. Chinese workers in particular provided cheap labor for low-wage sectors such as shoes and cigars. More significant, however, was the systematic application of skilled labor, which San Francisco had in abundance. California's combination of talent and economic opportunity (especially for white men) made for a flurry of entrepreneurial activity and technical innovation. Clear evidence of this can be found in the number of new firms in the city and in the welter of innovative products, from blue jeans to hydraulic nozzles. Money and creativity unloosed the genie of growth, allowing industrialization despite high wages. With more than 350,000 residents by 1900, the city held one-fifth of the populace of the entire West Coast and had become the seventh-largest city in the country.[8]

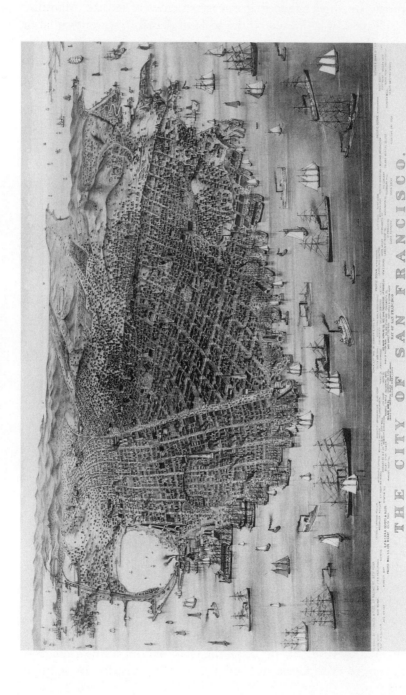

FIGURE 6.2. The City of San Francisco, Bird's Eye View, c. 1880.
Source: Lithograph by Currier and Ives. Reproduced courtesy of the Bancroft Library, University of California, Berkeley.

The industrialization of San Francisco in the second half of the nineteenth century was based on resource processing, and on products not far removed from the basic inputs of nature. Such was the technology of the time and the reality of California's resource-rich economy.[9] The four dominant sectors illustrate this: lumber and woodworking (lumber yards, planing mills, furniture); metalworking and machining (iron foundries, smelters, brass casting, metal plating, machine making); food processing (sugar milling, canning, coffee milling, biscuits and flour, alcoholic beverages); and animal processing (butchering, tanning, tallow, soap, glue, and candles).[10] Smaller sectors were not far removed from natural materials, either: garments made of wool, leather, and cotton (also blankets, shoes, and harnesses for horses), luxuries made of gold, silver, and tobacco (jewelry, cigars), paints and oils made of natural oils and pigments, energy supplies (coal, coal-gas, wood, methane, petroleum, and electricity). Even the most sophisticated products of the time, such as houses, mining machinery, carriages, and ships, were made with very humble materials: cast iron, brass, wood, tung-oil paints, iron nails, and so forth. Printing and publishing might be considered the furthest removed but this involved working with large quantities of paper and cloth. Without being overly reductionist—since the essence of things such as reapers, pianos, and ladies' dresses lies more in design and skill than in materials—it is nonetheless vital to recognize how close to the bone these early industries were. As technology advanced into new types of goods and materials, it would leapfrog over San Francisco's humble manufacturing base.

The Industrial Landscape of San Francisco, 1875–1905

Within San Francisco's expansive city limits, industry spread out rapidly after the Civil War.[11] Three patterns emerged in the period up to the turn of the century. First, industry was positioned along the waterfront, within a few blocks of the bay, in a great crescent from North Beach to Hunter's Point. Although this was a reasonably continuous strip, it can usefully be divided into six segments: North Beach, Downtown, Rincon Hill, Mission Creek, Potrero Hill, and Islais Creek. Second, the bulk of industry headed to the South of Market, far beyond the bounds of the Gold Rush city (where downtown offices, banks, and Chinatown remained), stretching the city southward for several miles. Third, while factories and warehouses of various kinds stood cheek by jowl along the waterfront, there was a discernible degree of areal specialization by industrial sector: machining around Rincon Hill, foodstuffs north of Market, luxuries downtown, lumber and wood products along Mission Creek, and animal products at Islais Creek (see Figure 6.3).[12] On the other hand, two of the most common land uses—energy and wine warehousing—could be found almost everywhere along the waterfront, for easy shipping (ships also used great quantities of fuel from these depots).

The area north of Market Street was characterized early on by a central cluster of consumer-goods manufacturers that arose in the old part of the city. Along the waterfront were several food merchants and processors such as Ghirardelli's Coffee, Pacific Coast Syrup, National and Capital Flour, National Biscuit, Sacramento Biscuit, Shilling Spices, and Tillman-Bendels (coffee and spices). A few clothing merchant-manufacturers, such as Levi Strauss, and boot and shoe makers could be found there, too. In the decade after the Civil War, 200 cigar makers clustered near Chinatown, relying on the cheap labor of thousands of Chinese men released from railroad building and mining, where they lived mostly within the restricted boundaries of Chinatown.

To the west, up Market Street, women's and children's clothes were made and sold. Around Union Square could be found specialty clothiers, jewelers and silversmiths, and makers of luxury goods such as pianos and billiard tables. This was part of the separation of the main shopping district from the merchant-financial cluster along Montgomery Street. Centered around Market and Third streets, where the three largest newspapers were located by the 1890s, was a flourishing printing and publishing industry, with more than 100 small printers, Bancroft's Book Emporium, and magazines and broadsheets galore.[13]

On the city's northern edge a substantial industrial zone arose in North Beach. This zone stretched from Black Point to the northwest around Telegraph Hill to the foot of Broadway on its southeast flank. Its first big factory was the Pioneer Woolen Mills (later Ghirardelli Chocolate and now Ghirardelli Square shopping center). It included a number of food-processing operations, such as sugar refineries (Bay—later American—Sugar and Western Sugar), canneries (Presidio, King Morse, and later the merged California Fruit Canners Association, plus a can factory), breweries and malt factories (St. Louis, North Beach, Empire), a flour mill (Globe), a wine warehouse, biscuit makers (National and American) and even a mustard mill (CC Burr). At the same time, North Beach had some key companies in the metals trade: Selby's lead and silver smelter, Andrew Haladie's wire works, Joshua Hendy's machine shop, and Vulcan Iron Works. And there were lumber yards, sawmills, and a sulfur refinery as well.

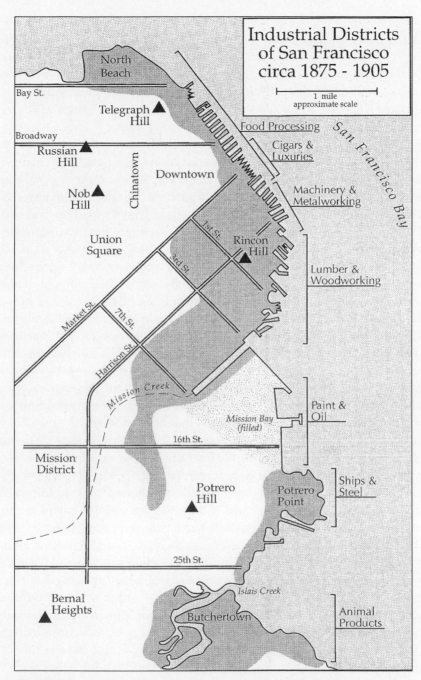

FIGURE 6.3. Industrial Districts of San Francisco, c. 1875–1905.
Drawn by Darin Jensen.

The South of Market industrial zone was anchored by the Rincon Hill district, from Market Street to South Beach (China Cove) and west to Third Street. This area was dominated by the city's major machinery and metalworking cluster, made up of foundries, plating shops, iron works, and machine shops. Of these, Union Iron Works was the most famous and occupied a full city block, but there were dozens of others, such as Risdon Ironworks, Pioneer Ironworks, Tatum and Bowen, Eureka Foundry, National Ironworks, Golden West Plating, Roylance Brass, Montague Tinware, California Engineering Works, Union Gas Engine, Oriental Gas Engine, Bryan Elevator, Selby Smelter, Judson Iron and Steel, John Finn Metal, Pacific Hardware and Steel, and so forth. Mining equipment, such as drills, rock crushers, steam engines, hoists, derricks, pumps, and nozzles, was the main focus of this complex; San Francisco revolutionized mining with this equipment, which it shipped throughout the West and around the world.[14] From there the district branched into ship engines, flour mills, fruit presses, sugar mills, sawmills, and farm equipment. Then, too, local companies turned out such necessities of urban life as lead stoves, pipe, locks and safes, bolts and boilers. There were, in addition, a number of coal yards and gasworks, lumber yards on the eastern and southern flanks of Rincon Hill, several wine warehouses, planing mills, a flour mill, a cooperage, and the like. By the turn of the century, Rincon Hill had, if anything, become more specialized in metals, despite the loss of some companies and the addition of Folger's Coffee roasting plant and Shilling Spices (which had both moved south across Market Street).

Southwest of Rincon Hill was another industrial district along Mission Creek, a mile and more south of Downtown. Mission Creek had been reduced to a finger canal for ship access with the filling of most of Mission Bay in the Civil War decade, and the Southern Pacific rail lines lay parallel, a block north of the canal (between King and Townsend, terminating at 3rd Street, site of Southern Pacific's headquarters before 1906). The companies operating there were a mixed lot, but the concentration of lumber and woodworking is significant. Lumber was shipped from the northern coast of California to San Francisco, where it arrived at huge depots along Mission Creek such as Hooper's, Simpson, Sierra, California, and Truckee Lumber (lumber wharves east of Rincon Hill had mostly disappeared by 1905, probably because the industry had consolidated). From there it was distributed to smaller wood yards and planing mills to be cut to size or shaped into parts and fixtures. Planing mills, such as Stockton, Usona, Progressive, Moore, and Mechanics, were numerous in the Mission Creek district (with a few left on Rincon Hill and in North

Beach). As the chief building material of the day (and thus the largest single industrial sector in the state up to 1910), wood went into any number of products. In the Mission Creek area it went into boxes, barrels, matches, furniture, carriages, machines, and ships. Furniture makers included Becker & Dillman, Frei, San Francisco, Emanuel's West Coast, Cordes, M. Friedman, and George Fuller. Carriage and wagon manufacturers were fewer but notable: Studebaker Brothers, Pacific, Waterhouse & Lister, Union, and Hammond Car Works (and Dunham Carriage, still over on Rincon Hill). There were two small shipyards: Hay's and Boole and Beaton. Mission Creek also had its share of foodstuffs factories and warehouses. It was the original site of such large operations as Claus Spreckels' (California) Sugar, Miller & Lux Butchers, and A. Lusk Canning, before they moved elsewhere, and later of F. Cutting Canning, Yosemite Flour, Chicago, Jackson and Milwaukee breweries, South Park Malt, Code-Portwood Canning, Philadelphia Vinegar, and Eureka Wine.[15] There were a number of metalworking facilities, among them American Can, Francis Smith Steel Pipe, Acme Brass, Brunig's Machine, Garnett Brass (and the Excelsior Smelter, right next to Starr King Primary School!). And, finally, there were such odds and ends as gasworks, wool warehouses, marble and stone yards, a cigar maker, a glass works, and a broom factory.

Many factories moved farther out in the 1870s and 1880s, south of Mission Creek (roughly 14th Street) in an area that was not platted until after the Civil War. This consisted of two districts, Potrero Hill and, south of 25th Street, Islais Creek. A distinctive group of mostly noxious and space-extensive activities developed in these outlying areas: animal processing, chemicals (paints and oils), and shipbuilding. (Powder works were another noxious industry already located far outside the built-up part of the city, near the Pacific Ocean—these are not shown in Figure 6.3).

A notable cluster of heavy industry took shape on Potrero Point, more than two miles south of Downtown on a rocky promontory across what was formerly Mission Bay. The region's first substantial steel producer, Pacific Rolling Mills, went up in 1868 after gaining a grant of tidewater lands from the state. Union Iron Works relocated its shipyards there in 1878, joined in the 1880s by John Myers Shipyard and in the 1890s by Risdon Iron Locomotive Works and shipyard. Other major factories at the Point were Tubbs' Cordage, Spreckels' California (later Western) sugar refinery (the largest factory on the West Coast in value of output), San Francisco Gas (and Electric) Works, and Arctic Oil Works (when San Francisco was the biggest whaling port in the world, circa 1882–1908), replaced by Union Oil's petroleum depot by the turn of the century.[16]

On the northwest flank of Potrero Hill a tongue of industry moved along upper Mission Creek—into what would later be known as the Northern Mission district—around 16th and Potrero streets. In the 1860s boats could still make their way up the creek, and the railroad followed it down toward the San Mateo peninsula (Mission Creek would soon disappear into a culvert beneath Harrison Street, but its waters were used by factories until the middle of the twentieth century). The character of this group of factories was dominated by animal products. These included Roth, Blum pork packing, two tanneries, Worden Grease and Varnish, San Francisco Candle, John Counihan Tannery, SF Glue Works, Golden Gate Woolen mills, and several soap manufacturers. Another distinctive segment was chemicals, which at the time meant chiefly paints and pigments: Whittier-Fuller moved out from Rincon Hill to the Potrero district by the turn of the century and was joined by Stauffer's Potrero works. Finally, there was a terra cotta tile works, a trunk factory, a mattress factory, a glass works, two breweries, and a wine company. The biggest single factory to come into this area by 1905 was the American Steel and Wire/California Wire Cloth company operation at 16th and Harrison. Farther out, where open lands remained, there were a couple of commercial nurseries and a dairy.

Below 25th Street, three miles south of Downtown, lay the Butchertown district around Islais Creek. On the initiative of wholesale cattlemen and packers, led by Henry Miller, the state ceded waterfront lands for a Butcher's Preservation (an early form of industrial park), built on fill on the south shore at the mouth of Islais Creek in the 1880s. Most of the city's butchers relocated there to be free of nuisance complaints and dump their offal into the bay. The district around Butchertown, on both sides of Islais Creek, eventually included several cattle yards, twenty-three wholesale butchers, five tanneries, a wool pullery, two packing houses, a glue works, and tallow plants—many owned by the vertically integrated Miller & Lux, the largest agribusiness enterprise of the nineteenth century. A notable accompaniment to animal processing was a budding fertilizer sector, including Bryle LaCoste and Pacific Guano. Finally, out at Hunter's Point, beyond Butchertown, was the big California Drydock, Neylan's Shipyard, California Fireworks Company, Albion Brewery, and a couple of orphan asylums.[17]

The post–Civil War flow of industry split the city in two: an industrial working class South of Market and a residential bourgeois northern tier and, to a lesser extent, an east-west schism as well. The well-to-do who had occupied Rincon Hill around South Park in the 1850s fled back north of Market, riding cable cars (introduced by Hallidie in the late 1870s) to the safety of Nob Hill. Rich men who had sought suburban refuge in the early Mission district, such as James Phelan Sr. and John Spreckels, abandoned it once it became *declassé*. Thereafter, the bourgeoisie settled a broad strip across the northern tier of the city, out to Van Ness Avenue and across into Pacific Heights and the Western Addition.

North of Market, the main redoubts of the working class were Chinatown (much reduced by the Chinese Exclusion Acts of 1882 and 1892) and North Beach, which became mostly Italian by the turn of the century. Telegraph Hill was built up solidly only after the great earthquake and fire leveled the whole area. When the Marina district opened up to the west after the 1915 Panama-Pacific Exhibition was vacated, better-off workers and middle-class Italians moved out there. Sailors and longshoremen occupied a narrow strip of boarding houses and hotels along the waterfront on the eastern edge of the city from Telegraph Hill to Rincon Hill, which became part of the so-called Barbary Coast (centered just south of Broadway) (see Figure 6.3).

Meanwhile, the South of Market became the principal area of working-class housing in the city, because it held the core industrial zones around Rincon Hill and along Mission Creek.[18] Because workers usually walked to work, roughly four-fifths of San Francisco's laborers in the late nineteenth century lived within one-half mile of their jobs. This meant that working-class residences were built in close proximity to the industrial strip running along the waterfront. The densest area of settlement piled up below Market Street down to Harrison, from the waterfront west to 12th Street. This was very likely the densest residential district west of Manhattan before the earthquake and fire, with many workers living in group lodgings, boarding houses, or multistory flats. The in-fill of housing throughout this zone was virtually complete, and housing grew up side by side with industrial sites. Most of the residents in this area were Irish immigrants. The 1906 earthquake almost certainly killed thousands of people South of Market who were never accounted for, but the numbers were hushed up in the aftermath so as not to unduly disturb the business revival of the city.

Somewhat more prosperous workers could take the cable car or horse-car jitneys to upper Market Street or out to the Mission district, which was also filling up at this time; by 1900 there was a built-out corridor a couple of blocks on either side of Mission Street as far as 22nd Street. These better-paid employees, mostly Anglo-Scots and Germans, could buy or build single-family homes, either from small builders or as lumberyard kits. At the same time, extensions of the industrial zone southward propelled the outward flow of working-class settlement, and working-class neigh-borhoods began dotting Potrero Hill, the former Mission Creek lowlands, and even Bernal Heights to the south (see Figure 6.3). Specific patches of housing grew up in direct juxtaposition to the pioneer outlying factory dis-tricts, such as the mean quarters of Dogtown just behind Potrero Point, around Butchertown (in a district then called "South San Francisco"), and next to the city hospital on the western flank of Potrero Hill. After the earthquake and fire, the rest of the Mission district filled in, Bernal Heights and Potrero Hill became covered with houses, and the working-class realms pushed farther south into the newly platted Excelsior, Ingleside, Outer Mission, and Portola districts. Many of the same capitalists back-ing industrial expansion, such as the Crockers and Newhalls, were busy investing in land development in these outer lands.[19]

Last Gasp of the West Bay: South City and the Eclipse of San Francisco

A telling case of industrial decentralization is South San Francisco, ten miles away from downtown San Francisco and across the line into San Mateo County (see Figure 6.4). South City, as it came to be called, pro-vides a clear example of the forces of industrial restructuring, property ownership, and political maneuvering on industrial suburbanization. It was also the last stronghold of industry along the western shore of San Francisco Bay, which would be cut off by the burghers of San Mateo from continuing its southern march. Industrialization in the first half of the twentieth century would shift decisively to Oakland and the East Bay.

The area at the foot of Mount San Bruno was for many years the prop-erty of Charles Lux and was used as a staging area for cattle going to slaughter. After Lux died in 1887, his heirs and former partner, Henry Miller, cut a deal with Joseph Swift of Chicago, fronting for the Beef Trust. Swift's Western Meat Company factory went up on the Lux property in 1894. Western Meat represented the success of the Chicago packers in revolutionizing beef slaughter and packing; by means of improved mass production methods and unskilled labor, they could under-price local butchers. But the latter were able to hold off their new competitor for a decade by convincing San Franciscans that industrial meat was tainted. Swift had to orchestrate the entry of the new methods carefully, seeking a local partner in Miller & Lux and a strategic location just outside the city limits. Victory did not go to the swift until the 1906 earthquake lev-eled Butchertown and people had to try Swift's product. They found it

FIGURE 6.4. San Francisco Bay Manufacturing and Shipping District, c. 1925.

Source: San Francisco Commercial News. Reproduced courtesy of the Bancroft Library, University of California, Berkeley.

palatable, and soon several other packers and a new Union Stockyards came to South City. Ironically, Miller & Lux could not compete with the Chicago boys, and went into decline.[20]

Other factories were attracted to South City's emergent industrial district in the 1890s and 1900s, including an immense Steiger Terra Cotta & Pottery factory, a Whittier-Fuller paint factory, and Columbia Steel's mill, all the largest of their kind on the West Coast. Along with these came another half-dozen iron and steel plants supplying regional shipyards, builders, and metal shops. Housing again followed the outward march of manufacturing, and the valley south of Mount San Bruno quickly filled up with the little homes of the working class—who could neither commute from San Francisco nor afford homes in the elite cities of San Mateo County. South San Francisco remains resolutely proletarian to this day. Politically, South City was the creature of the corporations. Of the first four mayors of the town after it incorporated in 1908, two had worked for Western Meat, one for Southern Pacific, and one as a local realtor.[21]

Industrialization might have continued down the peninsula, but it ran into the border guards of the wealthy. San Mateo County had long been the rural redoubt of San Francisco's biggest capitalists, among them Darius Mills, Alvinius Hayward, and William Sharon; and the children of the barons, led by Francis Newlands and William Crocker, opened an exclusive residential enclave at Burlingame (Hillsborough) in the 1890s. Fearing for their sylvan landscapes, the grandees blocked a planned ASARCO copper-smelting plant in 1908 and closed off further expansion of South City. This gave added impetus to heavy industry to take the path of least resistance over to the East Bay. San Mateo's industrialization slowed down before and after World War I. Only the modest nodes at Redwood City and South San Francisco remained (barely visible in the frame of Figure 6.4).[22]

Of course, San Francisco's industrial base was still substantial through the first half of the twentieth century (see Figure 6.1). Factories and warehouses were rebuilt along with the rest of the city after the earthquake. While North Beach and the South of Market area remained industrial, little was left in the downtown area. A notable addition to Rincon Hill were the big coffee roasting and canning operations of Hills Brothers (inventors of the first vacuum-packed coffee) and MJB (Chase & Sanborn would locate at Potrero Hill).[23] South of Market also became the domain of scores of printing shops. There was an expansion toward the open areas south of Mission Creek as new factories and warehouses were added. The Mission district and the flanks of Potrero Hill filled up steadily with new plants such as McClintock-Marshall Steel, Pelton Water Wheel, Crown

Shirts, Remler Radio, Baker & Hamilton Machinery, Pacific Felt, Hyman
Shoes, Hellmann's Mayonnaise, Best Foods, Lyons-Magnus Fruit Products, Rainier, Regal-Pale, Hamm's Breweries, People's Bread, Orowheat
Baking, and John Cleese Mattresses, along with planing mills, industrial
laundries, dye works, and so on.[24] Some of the older companies remained,
like Sterling Mattress and Illinois-Pacific Glass, while others relocated
from more northern locations, Korbel Box, American Can, and Enterprise
Foundry among them.

The Islais Creek basin also became completely covered with large factories and warehouses such as Planter's Peanuts. By mid-century Islais
Creek was the largest industrial district in San Francisco. Hunter's Point
was taken over by the navy during World War II for its principal West
Coast dry dock operation. Meanwhile, as San Francisco County was built
out with housing during the interwar period, the southern and eastern parts
of the city kept their working-class character intact, even as African Americans were added at Hunter's Point–Bayview, Filipinos came into the South
of Market, Central Americans entered the Mission, and European ethnic-
ities blended into a generic category of white Americans.

But San Francisco's days of unrivaled glory were over, and it fell from
its perch atop the western states after the turn of the century. Los Angeles began its remarkable ascent, becoming the largest urban center in the
West by 1910, while Seattle, Portland, and Denver all grew substantially,
challenging San Francisco's commercial hegemony. This fall from grace
was paralleled by the fate of the city's industrial base. Manufacturing lev-
eled out from 1890 to 1930. While output continued to rise, employment
growth rates fell off for twenty years, revived in the World War I, then
slid again before collapsing in the Great Depression (Figure 6.1). The
city's saving grace was the banking sector, which held on to its position as
the financial heart of the Pacific Coast.[25]

San Francisco's industrial woes are often blamed on the earthquake and
fire of 1906, when people and businesses fled in droves to the suburbs. Permanent losses included Joshua Hendy's relocation to Sunnyvale in Santa
Clara County, Judson Steel to Emeryville, and Vulcan Iron Works to Oakland, and most of the wine merchants, who lost millions of gallons in the
quake. Yet most refugees returned as the city was reconstructed. A closer
look shows that San Francisco manufacturing was already in trouble before
the great quake, slow to rebound from the downturn of the 1890s. Any
number of companies, including Lusk Canning, Alex Hays Shipyard,
Excelsior Smelting, and Becker & Dillman Furniture, had already disappeared by 1905. Meanwhile, most new factories in the Bay Area were

being established in industrial suburbs like South San Francisco and Richmond by the turn of the century. These included a wide cross-section of industrial sectors—blasting powder, paint, chemicals, meatpacking, oil refining, sugar, lumber, machining, canning, steel, ships, vehicles, and electrical machinery. The East Bay was the principal beneficiary of this metropolitan industrialization.

Alameda and Contra Costa Counties together passed San Francisco (and the West Bay as a whole) in manufacturing output by 1910 and in employment by 1920 (see Figure 6.1). In a reflection of this shift, the port of San Francisco fell below one-half of all Bay Area tonnage handled by 1920. As a result, the Bay Area as a whole performed better than most historians acknowledge. From 1900 to 1920 the five-county population grew by 87 percent, employment by 163 percent, and value added by 617 percent (San Francisco, Alameda, Contra Costa, San Mateo, and Marin). From 1869 to 1935 the city region outgrew Philadelphia, Boston, and Baltimore in population, ran even with New York and Pittsburgh, and lost ground only to the industrial *Wunderkinder* of the East, Detroit and Cleveland, and to Los Angeles. Overall, in the years 1869 to 1935, the Bay Area improved its rank among U.S. metropolitan areas in total employment from seventeenth to sixteenth and in value-added from fifteenth to tenth. Bay shipping led the West as late as 1938, second only to New York in value (Los Angeles was still a distant sixth). It is important, therefore, not to confuse the fate of San Francisco with that of the metropolitan region as a whole. Industrial dispersal had much to do with their differing fates.[26]

Nonetheless, we still require an explanation for the tilt away from San Francisco toward the East Bay. One reason was simple space: as industry expanded and the scale of factories increased, there was pressure to seek large, cheap, accessible sites at the urban fringe. This was true of nurseries as well as of oil depots and refineries. Another reason was the impulse to follow the general march of the city and local markets outward, as in the case of lumberyards, breweries, tile works, and foundries, the last making civic infrastructure like manhole covers. A third was to avoid conflict with residential areas, as we have already seen. But certain overarching causes have been proposed.

One unifying explanation that had to be discarded long ago was that local companies fell prey to buyouts by eastern corporations after the turn of the century. Some large companies, such as Studebaker, Bethlehem Steel, and the Chicago Packers, did build or buy out local factories. But they added to the local mix rather than subtracting from it, and they never dominated California industry, which continued to proliferate its own firms and products. This theory was part of the mid-century adulation of the large corporation and Fordist mass production—which has fallen out of favor with the late twentieth-century success of California industry, especially Silicon Valley.[27]

Another popular candidate for a unitary explanation is high wages and labor organization. The militancy of San Francisco's workers from 1901 to 1918 is legendary, and electoral control of the city lay with the Union Labor Party from 1901 to 1911. There is some evidence that this induced capital to flee: Columbia Steel, for example, built a mill in Contra Costa County in part because, according to financier Joseph Grant, "labor conditions there would be less disturbed than in San Francisco." William Gerstle, president of the Chamber of Commerce, said in 1910 that in San Francisco "the cost of manufacturing is so high that we cannot compete with neighboring communities. Everything is on a competitive basis except labor, and this is due to the fact that we have not had the courage in San Francisco to enforce the open shop principle which prevails in our competitive cities." Yet wage rates alone do not set the terms for capitalist development, and low wages are not always an advantage. In fact, very high average wages compared to the rest of the United States never choked off growth in nineteenth-century San Francisco. In fact, they helped stimulate mass in-migration and consumption, and attracted the skilled labor conducive to innovation—all of which contributed to growth. So what had changed? If there is one area in which labor militancy had a clear effect, it was the reduction of low-wage Chinese labor; this probably lies behind the precipitous decline of such sectors as cigars, boots and shoes, blankets, and apparel in San Francisco before 1900.[28]

A third sweeping theory is the failure of a weak bourgeoisie to adequately promote the city's interests vis-à-vis its urban rivals. True, San Francisco capitalists were notoriously schismatic, as in the long feud between the Spreckelses and DeYoungs, and no doubt some ground was lost to better organized business classes in Los Angeles and Oakland.[29] The stillborn proposal by the Chamber of Commerce for an Islais Creek industrial park after the fire is one example of a bourgeois failure of will. Nonetheless, San Francisco's upper-class leaders were capable of joint efforts in many respects. James Phelan and his circle of Progressives were instrumental in reforming the city charter at the turn of the century. Following that, the earthquake forced businessmen to put aside differences, as did the struggle against organized labor. The major business associations merged under the Chamber of Commerce, elected merchant James Rolph mayor in 1911, and established a think tank that generated another city charter reform. Business engaged labor in repeated battles, finally defeating the unions citywide in 1921. A spectacular new civic center was erected between the world wars.[30] Overall, the problem was not so much that San Francisco's capitalists were weak as that they had a wide-ranging view of the field of investment—one that took in the whole Bay Area and most of the western United States besides (city capitalists were instrumental in building up Los Angeles, San Diego, California's interior, and even the Pacific Northwest). They were not simply advocates within the city limits.

But the most important single factor in the decentralization of manufacturing in the Bay Area was none of the above causes. It was industrialization itself, i.e., the force of technical and market change unleashed by capitalist accumulation. Swiftly growing sectors can erupt in quite unexpected venues, while stagnant sectors and established centers of industry fade away. The industrial base of capitalism has shifted repeatedly from era to era, recasting urban geography along the way.[31] San Francisco had led the way during the expansive stage in several early California industries, such as mining equipment, men's clothing and blankets, lumber and woodworking, iron work, carriages, foodstuffs, alcoholic beverages, paint, and animal byproducts. But three things happened to undermine the importance of the city's manufactures. The first is that some of its best products ceased to be very important in the mix of California industry in the twentieth century; these included mining equipment, leather harnesses, wooden barrels, whale baleen, steam boilers, cast iron stoves, buggy whips, leather belting, and Victorian house detail work. Second, conversely, most of the key industries of the twentieth century grew up almost entirely outside San Francisco, among them petroleum refining, alloy steel, automobiles, and chemicals. Third, in several sectors where the city had a foothold, the technical nature of the product or process changed so much as to become unrecognizable: guano gave way to ammonium nitrate fertilizer made through electrolysis; carriages and wagons turned into cars and trucks; coal was replaced by fuel oil, coal gas by methane, gas by electricity.[32]

By 1919 San Francisco was still prominent in only a few industrial sectors—printing and publishing, coffee and spices, chocolate, cigars, bags, and ship repair—none of them among California's leading industries of the day. Output in furniture, meatpacking, clothing, and machining, while still noteworthy, was fading. Beer and spirits were eliminated by Prohibition. Lumber, sugar, and canning had virtually disappeared. And, of course, the process of product and technical change kept moving forward, even where San Francisco knew success in the new century: Ford's Model T factory on 16th Street, opened in 1913, was replaced by a Model A factory in Richmond in 1931 as Ford responded to the competitive challenge of General Motors. The city continued to house many of the corporate headquarters of businesses involved in the new industries, such as Standard Oil, Zellerbach Paper, and Castle & Cooke foods, but that was due to its importance in finance and administration, not production.

Oakland Rising: The Industrialization of Alameda County

The forward wave of regional growth had shifted by 1900 to Oakland and the greater East Bay. Manufacturing in Alameda and Contra Costa Counties together surpassed San Francisco and the West Bay by World War I. The rapid acceleration of East Bay urbanization that went along with this industrial surge would create the greater Bay Area metropolis of the twentieth century (Figures 6.1 and 6.4).

Oakland began as one of several small towns around the bay, with the usual smattering of resource industries. By 1869 it was home to sixteen factories, including sawmills, tanneries, slaughterhouses, dairies, a jute mill, a flour mill, dry docks, and a brewery (the only thing out of the ordinary was a boot and shoe maker). The big turning point was the arrival of the Central Pacific in that year, after which population climbed from 10,500 to 35,000 in a decade—making Oakland the second city in the western United States for a generation. Although the rail terminus was officially San Francisco (trains were ferried across the bay from the 7th Street mole), the rail yards were a major employer in West Oakland and an attraction for manufacturers seeking access to California markets. Oakland industrialized rapidly through the rest of the century. Factories became abundant in the 1870s and the 1880s saw another thirty establishments spring up. By 1890 California Cotton was the largest cloth mill in the West, Josiah Lusk the biggest cannery, Pacific Coast Borax the largest producer of cleanser, and Lowell Manufacturing the biggest carriage works. But the best was yet to come, and Oakland still looked more like a satellite of San Francisco's diversified manufacturing complex than a realm of its own.[33]

The principal axes of Oakland's industrial belt ran along the waterfront and were reinforced by the rail line coming from the east along the estuary all the way to Oakland Point (1869), where it met a second line arriving from the north (1873). The principal manufacturing node lay within the original Oakland city grid, which ran from the waterfront to 12th Street, with Broadway as the central thoroughfare. Machining and woodworking were fixtures of the central district. A second cluster appeared a mile and a half west, in an area platted and annexed in the 1860s. The defining activity of West Oakland took place in the rail yards, but beyond them at Oakland Point (now long gone owing to surrounding fill) lay such space-extensive functions as lumberyards, shipyards, stockyards, tanneries, and slaughterhouses. A third cluster could be found two to three miles east across Lake Merritt in the Brooklyn and Fruit Vale districts (settled in the 1850s and annexed in 1872), which featured the Cal Cotton mill, a boot and shoe factory, saw and planing mills, early canneries, a flour mill, a pottery factory, and tanneries. A fourth formed in the 1880s and 1890s two-and-a-half miles north of Downtown along Temescal Creek, beyond the city limits, in what would later become Emeryville (incorporated 1899) and North Oakland (annexed 1897). Some of the largest factories in town, such as Judson Steel and Lowell Manufacturing, moved to the Emery district, while up the creek the principal site was the J. Lusk cannery, with 400 acres of fruit and vegetable gardens in the Temescal district. (Pacific Coast Borax and N. Clark & Sons Brick Works were across the Oakland estuary at Alameda Point, and there was another small industrial settlement at Oceanview, now West Berkeley, to the far north.)

The initial locus of housing was the 1850 city grid, with small settlements around workplaces in Brooklyn (Clinton and San Antonio districts) and Fruit Vale to the east. A few wealthy burghers like James DeFremery and George Pardee moved into the wide open spaces to the west, but after 1869 that area blossomed as a working-class district and the rich were displaced to the northeast around Lake Merritt. West Oakland filled in rapidly all the way up to 50th Street (roughly Temescal Creek), which gave the city a lopsided appearance for decades. Workers commuted on foot from the flatlands east of the industrial belt or rode the streetcars fanning out from downtown and West Oakland in a manner still clearly visible in the diagonals overlaying the regular street grids. The little industrial communities of Brooklyn—which grew less rapidly—retained their distinct identities long after incorporation, each with its own street grid, retail commerce, and ethnic flavor. For example, Jingletown, where Jack London grew up, still looks like any eastern mill village with an ethnic Catholic working class, though it has gone from Irish to Portuguese to Mexican and was bisected by a freeway in the 1940s.

The bay made Oakland a twin city rather than a suburb, but it was not a strong antipode to San Francisco in the nineteenth century. While many manufacturing companies were locally owned, giving the city a potentially independent economic base, the town's leading burghers, Horace Carpentier and Samuel Merritt, were master land speculators rather than industrialists. The local bourgeoisie gave away the waterfront not once but twice to private owners (themselves!), who turned it over to the Central Pacific based in San Francisco (indeed, Carpentier, Oakland's first mayor, worked as a lawyer for the railroad before he took the money and ran to New York). It took the city fifty years to regain control of its harbor (1911). Stirrings of boosterism could be seen in the bold proposal for a cross-bay bridge as early as 1863, but San Francisco capitalists felt little threat and were happy to join in Oakland's growth by investing in such things as a cable car system, paint factory, and real estate promotions.[34]

The sea change came after the depression of 1893–95. Oakland and the East Bay began a meteoric ascent, becoming an early metropolitan "edge city." Oakland was one of the three fastest-growing cities in the United States between 1900 and 1930, jumping from 67,000 to 284,000 inhabitants, and development spilled over into the neighboring towns of San Leandro (1872), Berkeley (1878), Alameda (1884), and Emeryville (1899). The earthquake of 1906 doubled Alameda County's population and industry overnight, and there was another trebling of employment during the boom of World War I. Output continued to rise in the 1920s, although employment slipped (Figure 6.1). Oakland was no longer an annex to the metropolitan core but a distinctive industrial arena in full bloom (see Figure 6.5). Breakthroughs in transportation were important, of course: repossession of the waterfront from the Southern Pacific allowed the city to develop its own port facilities, and the arrival of the Santa Fe (circa 1900) and Western Pacific (circa 1910) lowered freight rates. The East Bay grew on water and rails, not trucks, well into the twentieth century. But the port and rail system grew to serve industry as much as, if not more than, the other way around.[35]

Fundamental to the industrialization of the East Bay were the emergence of new sectors and major reorientations in older ones. First among Alameda County's peacetime industries after 1900 was food processing, chiefly canning. The East Bay became the principal node in the Bay Area's largest industry, which led the nation in canning output from 1890 to 1940. California packers and canners introduced the first name brands in food, standardization of produce, and mass advertising in foodstuffs (Del Monte brand was dreamed up at the Lusk company). They later set up the world's most advanced marketing and contracting system, tied to the new supermarket chains (such as Oakland's Safeway). And they pioneered new methods and products, such as the canned olive (invented in Berkeley). A major organizational restructuring of canning took place as well, as the industry underwent a marked concentration. In 1899 a dozen companies merged into the California Fruit Canners Association, headquartered in Oakland; in 1917 this group expanded into the giant CalPak (Del Monte) corporation, the leading agribusiness firm for much of the twentieth

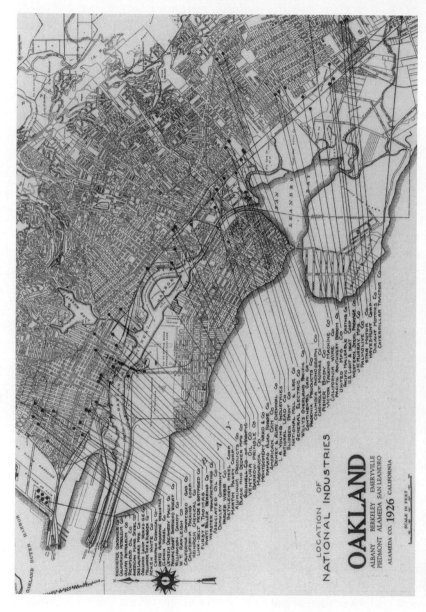

FIGURE 6.5. Location of National Industries, Central and East Oakland, c. 1926.
Source: Oakland Tribune Year Book. Reproduced courtesy of the Bancroft Library, University of California, Berkeley.

century (though its headquarters moved to San Francisco). A host of suppliers provided cans, jars, crates, and cartons to store and ship produce, as well as a stream of innovative machinery, such as pitters, peelers, and steamers. Many other food products were manufactured in the East Bay, including cereal, baby food, meat, and bread. Closely related were a dozen soap and cosmetics manufacturers. These factories were distributed along the length of the East Bay industrial belt and well into the outlying farming areas of southern Alameda County.[36]

Oakland's second leading sector was metalworking and machinery, which continued a long tradition in the Bay Area—but in a new era of steel alloys and high-speed cutting. Machine shops and foundries proliferated, clustering around Downtown and along the estuary-rail corridor. Oakland companies such as Union Machine Works, Bay City Iron, and Vulcan Foundry made machines for packaging, road grading, clothes washing, canning, and chemicals, as well as boilers, engines, turbines, and cast parts, some of which were unprecedented products. Upstream from metalworking was steel production, which finally developed as a significant industry in California in the twentieth century. The East Bay steel district, centered in Emeryville, was one of three that grew up around the Bay Area at the same time (the others were South San Francisco and Pittsburg, in Contra Costa County).[37]

The metal trades extended in several new directions. For a brief time, the Oakland estuary developed into an exceptional shipbuilding district, as companies transferred operations from San Francisco and wartime orders grew. Wooden shipbuilding first migrated to the estuary in the 1890s and was later joined by Moore and Scott (later Moore Drydock) and Bethlehem (moving most of Union Iron Works' former operations). During the peak years of World War I, a dozen shipbuilders employed 40,000 men, who put out 18 percent of U.S. production. Several companies supplied marine engines. The shift from San Francisco to Oakland appears to be tied to technical and product changes, such as steel construction, the Dreadnaught class of battleships, and oil tankers.[38]

No sector better exemplified the new age than automobiles and Fordist mass production, which swept into Oakland from Detroit in the teens. The city became host to more than fifty assembly and component plants in the interwar period. Chevrolet was first, in 1916, followed by Durant, Star, and Willys-Overland. Another pioneer was Coast Tire and Rubber in 1919, which was rapidly joined by a variety of tire and parts makers. Many of these were local companies, as were specialty assemblers like Fageol (buses) and Benjamin Holt (tractors). Within a generation, Holt's caterpillar tractors, invented in Stockton for the Sacramento Delta, had revolutionized

FIGURE 6.6. Chevrolet Factory, East Oakland, c. 1918.
Source: Photo by Bake Russell. Reproduced courtesy of the Bancroft Library,
University of California, Berkeley.

farming, warfare, and earth moving around the world. Automobile production filled in the vast expanse of East Oakland, after Chevy jumped 7.5
miles out to empty fields at Foothill and 70th Avenue (see Figure 6.6).[39]

The brand-new electrical machinery industry entered Oakland and the
East Bay in the 1910s with an influx of branch plants from General Electric, Westinghouse, Western Electric, and Victor, as well as local operations such as Marchand and Magnavox. These factories manufactured
lamps, motors, calculators, phonographs, and loudspeakers (invented in
Napa). Aircraft were a promising East Bay industry in the biplane era; some
thirty-five East Bay factories, including United Airlines and Standard Gas
Engine in Oakland and Jacuzzi and Hall-Scott Motors in Berkeley,
supplied airplane parts during World War I. For a time the new Oakland
airport, completed in 1926 at the eastern edge of the city, was the premier
airfield on the Pacific Coast.

The new wave of industrialization stretched the metropolitan area of
Alameda County dramatically north and east. Hand in hand with industry

came extensive residential development and land speculation. As west side industry built up, the flatlands of the north county up through Berkeley and Albany (1908) filled in, creating a sea of small working-class homes. During its period of rapid growth, from 1900 to 1930, the East Bay developed one of the most extensive streetcar networks in the country, owned by the Key System and Southern Pacific. Trolleys and good wages allowed considerable lateral mobility; so workers moved eastward all the way to the edge of the upper-class neighborhoods in the foothills. East Oakland beyond Fruitvale—largely vacant until World War I despite annexation in 1908—filled in rapidly during the 1920s. Subsequently, tracts such as Melrose Highlands, built by the Realty Syndicate, and Havenscourt, built by Wickham Havens, were developed expressly for workers at the new auto factories.[40]

The East Bay had its own striking examples of local political initiatives to steer development. One was the creation of Emeryville. At a stroke, an emerging satellite of Oakland became an independent city devoted wholly to industry—one of the first such entities in the United States, twelve years before the incorporation of South San Francisco. By 1935 little Emeryville (only 1.2 square miles) was home to more than 100 manufacturing plants. The town excluded all but a few hundred working-class residents and operated as a tax haven for industry. The manager of Judson Steel, Walter Christie, served as mayor for the first forty years of the town's existence, and was succeeded by Al LaCoste, a packinghouse boss, who ruled for the next three decades. But even reputable Berkeley put in a sophisticated zoning ordinance circa 1910 to protect factory owners from complaints by residential neighbors.[41]

By the turn of the century, Oakland was generating powerful capitalists willing to do battle with San Francisco over water supplies, port expansion, and industrial growth. A Greater Oakland movement got underway in 1896 to push for civic improvements, and the Oakland Chamber of Commerce campaigned tirelessly to attract investors. Francis Marion Smith was Oakland's first great booster capitalist, putting together the Key System of trolleys and building his Realty Syndicate into one of the biggest residential developers in the country (13,000 acres in 1900, almost 100 tracts complete by 1911). George Pardee, mayor 1893–95, went on to become governor of California, while Joseph Knowland became a powerful voice for local interests while serving six terms in Congress. The iconic figure of the new Oakland was nonpartisan mayor Frank Mott (1905–15), who brought several civic improvement plans to fruition. One was a skyscraper city hall that turned its backside to San Francisco. Another was the liberation of the port from Southern Pacific. Mott aggressively annexed all of East Oakland while it was still open land and tried forcibly to add Berkeley. Labor repression was something Oakland's burghers were especially good at, as when Mayor Pardee and Councilman Mott handed out pick handles to vigilantes who confronted Coxie's "army of the poor" in 1893.

By the 1920s Oakland was a major player in California politics. Joseph Knowland, who bought the city's main newspaper, the *Tribune*, became Oakland's chief powerbroker and the leading force in the state Republican Party for thirty years. He promoted Earl Warren to district attorney and then to governor of California from 1940 to 1954, and his son, William, to the U.S. Senate and to the position of Senate minority leader in the 1940s and 1950s. Henry Kaiser and Walter Bechtel built their construction empires out of Oakland in the 1920s and 1930s, partly on the strength of local projects such as the Alameda Tube, the Bay Bridge, and the Caldecott Tunnel. Kaiser led the "Six Companies" in building Boulder Dam, then became a figure of national importance in the Democratic Party by allying with Franklin Roosevelt (during World War II he became one of the world's largest employers, with roughly 250,000 workers in his shipyards, building sites, and factories).[42]

San Francisco's business leaders were alert to the challenge presented by Oakland to their hegemony over a burgeoning metropolis. Hoping to follow the lead of New York's metropolitan consolidation and L.A.'s aggressive annexations, James Phelan and his Progressive allies put together a Greater San Francisco Association to attempt political unification of the region. This plan went down to defeat in a statewide vote in 1912, against opposition led by Joe Knowland and Oakland's business community. Attempts to formalize a cooperative relationship under a Regional Plan Association started by Phelan in the 1920s also came to naught. Oakland's own attempts to annex Berkeley in 1908 and to merge city and county in the late 1920s failed just as miserably. Of course, the regional business class on both sides of the bay was acutely aware of the challenge presented by Los Angeles to the economic supremacy of the north, so some cooperation was possible. In the 1910s Oakland's leaders supported the Hetch Hetchy water plan, the Panama-Pacific Exhibition, and regional unification by bridge, interurban rail, and state highways. And during the Depression era, regional leaders were able to pull together on such infrastructural projects as the trans-bay bridges.[43]

All the same, San Francisco capitalists, undaunted by shifting industrial geography or political opposition, kept investing in an expansive metropolitan fringe around the Bay Area. In Oakland they were backers or owners of such firms as Parr Terminal, Moore-Scott Ships, and Hunt Brothers Canners. They invested in the East Bay's streetcar, rail, gas, and electric infrastructure. Industrial rivalries made little difference to financiers and realtors, who could play both sides of the table and hedge their bets. Oakland Bank of Savings merged with San Francisco's Mercantile Trust to form American Bank and Trust Company, the region's second largest bank, in 1921. Bank of Italy opened branches there, and Coldwell, Cornwall, and Banker Realty joined the rush in the 1920s. Several leading San Francisco businessmen, among them Wallace Alexander and Isaias Hellman, made their homes in Oakland's posh hills by the 1910s.

The Contra Costa Shore

The northern tier of the new East Bay industrial belt appeared from the 1870s to the 1920s in Contra Costa County along the banks of the Sacramento River. Contra Costa specialized in giant resource-intensive plants, processing explosives, chemicals, oil, sugar, cement, lumber, silver, lead, and steel. It leapt into the picture quite suddenly in the 1890s, and by 1900 the country's industrial output exceeded that of Alameda County (Figure 6.1). By 1906 some forty factories had opened along the river's south shore, including more than half a dozen of the largest factories of their kind in the country, C&H sugar, Standard Oil of California, Union Oil, Redwood Manufacturers, and Hercules powder among them. By 1920 the county's various docks carried more than half the tonnage on the bay, principally in petroleum. Edged out by Alameda County in the 1920s in value of output, Contra Costa did much better than its Bay Area rivals in the Depression and by 1940 was the second county in the state, after Los Angeles, in value of industrial output.[44]

Contra Costa developed a peculiar urban-industrial landscape of worker villages and company towns owing to the nature of its industries (see Figure 6.7). While some factories employed hundreds of workers, almost all were capital-intensive high-throughput operations that generated less total employment than the myriad workshops of Oakland and San Francisco (Figure 6.1). The most extreme form of this occurred at the several powder works, employing Chinese men who lived in bunkhouses. (Chinese were seen as expendable in the face of frequent explosions.) Such places as Hercules, Rodeo, and Cowell were little more than company towns. Crockett, the third-largest settlement in the county, was settled mostly by sugar workers. Pittsburg, the second largest, was mostly a steel town. Only Richmond, at the western end of the industrial belt and the terminus of the Santa Fe Railroad (1899), became a small city, counting eighty factories and 23,000 people by 1940. Contra Costa County's population came to only 32,000 in 1910 and a rather modest 99,000 by 1940—in sharp contrast to the rampant urbanization in Alameda County.[45]

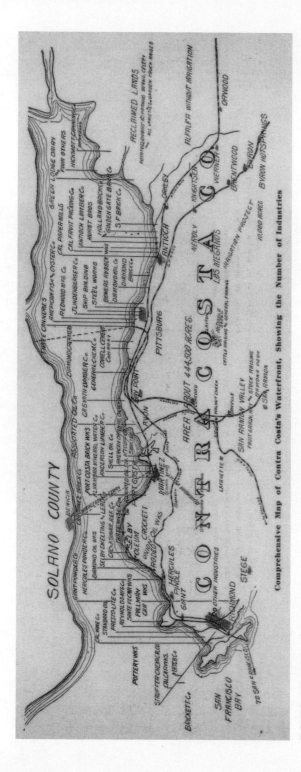

Comprehensive Map of Contra Costa's Waterfront, Showing the Number of Industries

FIGURE 6.7. Map of Contra Costa Waterfront, c. 1915.

Source: Chamber of Commerce pamphlet. Reproduced courtesy of the Bancroft Library, University of California, Berkeley.

The first manufactures to come to Contra Costa County were powder and dynamite works serving the mining industry. Atlas Powder and California Power Works moved out from San Francisco circa 1880 and were joined by half a dozen others thereafter. These were leftovers from the mining era, companies fleeing from nuisances complaints in the city. But a new industrial age was dawning, and it soon made its mark in Contra Costa. Chemical plants entered the picture at the turn of the century, as demand for sulfuric acid, chlorine, and ammonia fertilizers increased with advances in chemistry and industrial agriculture. Peyton Chemical was first, in 1898, followed by Stauffer Chemical in Stege (Richmond), Great Western Electro-Chemical, and others. Then came the oil refineries, another index of the new industrial era. A band of oil refineries along the river made the Bay Area one of the chief refining centers in the United States. The first big refinery was Union Oil in 1896; Standard Oil followed at Richmond in 1901, and four others came in soon thereafter. Oil came by pipeline, ship, and tank-car from the San Joaquin and Ventura fields.[46]

Another major industry was foodstuffs. In the nineteenth century the wheat boom had given birth to Port Costa, a rump town fronted by miles of docks for transshipment from rail to ship; the biggest warehouses went up circa 1880. But Port Costa and wheat went into sharp decline in the 1890s, when Contra Costa County was just catching fire. The fish-packing industry started early, too, but had more staying power. The great California fish-packing industry (famous from Monterey's Cannery Row) began along the Sacramento River: the first cannery to open was F. E. Booth, in 1875; three more plants were there by the early 1880s and seventeen still survived in 1940. A whaling station and rendering plant operated for many years at Point San Pablo (Richmond). California & Hawaiian (C&H) built an immense sugar mill at Carquinez Straits at the turn of the century, while the crenellated fortress of Winehaven, built by the California Wine Association at Point Molate (Richmond), was the biggest winery in the world before Prohibition closed it down.[47]

Primary metals were another mainstay of Contra Costa industry. Selby's lead smelter and shot works (later ASARCO) moved from San Francisco to the Carquinez Straits in 1884, adding gold and silver smelting and a cartridge factory. Copper smelting was first tried in 1864, but the most impressive operation came in the twentieth century with Mountain Copper (Mococo) at Bulls Head Point (Martinez). Steel came to the county in 1908 when Columbia Steel (later U.S. Steel) chose a site upriver at New York of the Pacific (and quickly changed the name to Pittsburg, misspelled by the locals). Also significant in the Contra Costa industrial complex were wood, paper, and building materials. Building materials went through a major restructuring around the turn of the century, with the introduction of Portland cement, better-quality sawmills, and large-scale use of asbestos. These technological changes featured in the shift of industry to Contra Costa: the Redwood Manufacturing Company's lumber mill, Cowell's cement plant, Johns Manville's asbestos works, and the California paper and cardboard mill.

San Francisco capitalists dominated the development of Contra Costa County, which was more an industrial colony than Oakland. Almost all the county's major factories, including Selby, Great Western Electro-Chemical, and Redwood Manufacturing Company, were dreamed up and financed within the city. San Francisco financiers orchestrated the rail, water, oil, and electricity networks that fed the new industrial district, including the crucial link to the Santa Fe that broke the Southern Pacific's rail monopoly into the Bay Area. The county's largest city, Richmond, was almost entirely a creature of San Francisco designs: Standard Oil of California was city based, plant manager William Rheem organized the local trolley line to tie Richmond to the rest of the East Bay, H. C. Cutting put together the company that developed the inner harbor, and Fred Parr came in to build the outer harbor terminals and to negotiate Ford's move from San Francisco to Richmond. Other Richmond enterprises, such as Stauffer, Winehaven, and Atlas, were funded by San Francisco investors. Oakland capital was represented by John Nicholl, who owned most of Point Richmond, where the Santa Fe terminus and rail yards were located.

Several upriver towns were also founded by city capitalists: Port Costa was the brainchild of merchant Isaac Friedlander, Pittsburg was engineered by Columbia Steel, and Crockett grew up under the aegis of C&H Sugar. Pullman's sleeping car works in Richmond was one of the few outside corporations to come into the area (Dupont, which bought out Grant Powder, was another). San Francisco's business leaders had a clear regional perspective and strategy for industrial decentralization, and they dominated the scene, given the paucity of local capital, unions, and working-class voters. There was little fractiousness from Contra Costa—unlike Oakland or even little Vallejo, just across the river in Solano County, which helped sink San Francisco's bid to be the home port for the U.S. Navy's Pacific Fleet. Contra Costa County was a South City or Emeryville writ large—a clean slate on which big capital could write its industrial narrative unimpeded.[48]

San Francisco's urban geography has been deeply shaped by industrialization, and industry has helped lead the outward march of the metropolis from the Civil War to this day. Suburbanization of industry appears to be the normal mode of urban growth. This can be seen in the spatial expansion of industry within the city of San Francisco, down the Peninsula, over to Oakland, and up along the Contra Costa shore. In fact, the tendency for industry to seek spacious quarters at the fringe is so marked from the outset that it makes no sense to speak of an "old industrial core" that later suburbanized. Indeed, San Francisco's original core of workshops had been exceeded in importance by the South of Market by 1870, and the latter by more southerly industrial districts by 1890. San Francisco as a whole was overtaken in the early twentieth century by the East Bay, which itself was spreading out rapidly from central Oakland to West Oakland, then to Emeryville and East Oakland, and along the Contra Costa shore from Richmond to Crockett to Antioch.

The reasons for new eruptions of industrial activity at the urban fringe are several. Land prices and speculative gains are one, as in the founding of Richmond around the new rail terminus. Better infrastructure is another, as improved rail access or harbor facilities opened up, as they did in Oakland. Less militant labor beyond the reach of the powerful San Francisco unions is a third, as in Contra Costa. Spatial politics enters with a vengeance in cases such as the Chicago meatpackers' avoiding the city's butchers or in the formation of Emeryville as an industrial enclave city. Most strikingly, spatial expansion has been closely associated with the opening up of new industrial sectors and the restructuring of old ones—the geography of the metropolis has been periodically (re)constituted along with its productive base.

Of course, there are always multiple and contingent forces at work, configuring parts of the urban complex differently. South City ended up a stunted-growth node, Contra Costa a string of industrial colonies, and Oakland a ferociously independent twin star to San Francisco. These particulars matter a great deal for the micro-geography of the Bay Area. But the metropolis was bound to expand geographically as long as its economy grew, and capital was destined to flow throughout the region as it developed. San Francisco investors, who led the way from the outset, had a metropolitan vision, and they planted the seeds for most of the development around the bay. As the editor of the *Chamber of Commerce Journal* wrote in 1912, "in San Francisco are made most of the great plans for state development," and he went on to quote H. C. Cutting, developer of Richmond's inner harbor, who said: "This growth means as much to San Francisco as though it took place within her own city limits, for financially it is all one. We grow together."[49]

7

Industrial Suburbs and the Growth of Metropolitan Pittsburgh, 1870-1920

EDWARD K. MULLER

W hen the field workers of the now famous Pittsburgh Survey arrived in Pittsburgh in the fall of 1907 to begin their investigations, they were already aware that the city was part of a much larger metropolitan district. From a compact urban area of fewer than 100,000 people in 1850, Pittsburgh had grown into a metropolitan area of more than a million. By 1920 the metropolitan region contained between 1.3 and 1.5 million inhabitants and sprawled thirty to forty miles from downtown into six counties. Even by the U.S. Census's more restrictive measure, Pittsburgh ranked fifth among all metropolitan districts in the nation.[1] Writing in the Survey's initial report of its findings in 1909, Robert A. Woods (head of South End House, a Boston settlement house) aptly noted that Pittsburgh's extensive metropolitan area resulted from widely dispersed industrial development, not simply residential expansion. "Pittsburgh has grown into an industrial metropolis with outlying manufacturing towns reaching along the rivers, and following course of all the railroads for a distance of thirty or forty miles. The time is

Reprinted with minor revisions from Edward K. Muller, "Industrial Suburbs and the Growth of Metropolitan Pittsburgh, 1870–1920," *Journal of Historical Geography* 27 (2001): 58–73, by permission of the publisher Academic Press.

soon coming when all the large industries will be eliminated from the city, and Pittsburgh proper will become simply the commercial and cultural headquarters of its district."[2] Residential dormitory suburbs contributed less than 10 percent of the total metropolitan population in 1920.[3]

Survey team members also recognized that the industrial activities of this metropolitan region were interconnected. Paul U. Kellogg, editor and mastermind of the Survey, described the physical linkages in his introduction to the Survey's reports: "Again, there is a temptation to define Pittsburgh in terms of the matrix in which the community is set, and the impress of this matrix on the soul of its people no less than on the senses of the visitor. Pipe lines that carry oil and gas, waterways that float an acreage of coal barges, four track rails worn bright with weighty ore cars, wires surcharged with a ruthless voltage or delicately sensitive to speech and codes, bind here a district of vast natural resources into one organic whole."[4] In addition to these largely visible transportation and communication links, capital, labor markets, and the division of labor in production processes bound the dispersed communities of this sprawling complex into a metropolitan region.

In this essay I argue that in the half century after 1870 tremendous growth in manufacturing and major reorganization of production under corporate capitalism drove the spatial extension of development far beyond the traditional urban core.[5] During the middle third of the nineteenth century, industrial expansion occurred incrementally at the city's edges with the establishment of new glass and iron plants. Beginning in the 1870s, however, large, integrated steel works sprang up well beyond these contiguous locations. Other capital-intensive mass-production plants in, for example, the electrical equipment and aluminum industries, which often evolved from smaller incubator origins in the city, joined the movement of steel firms out of the city. While these events manifested the workings of the product cycle, new firms with modern factories and small, traditionally organized companies also proliferated at these metropolitan locations.

Other forces contributed to the extensive sprawl of industry. Topographical barriers prevented simple incremental spatial expansion during this period of rapid growth, and widely scattered natural resources further attracted new industrial development. Speculation in real estate suitable for industry prepared sites and hastened development, while minimal governmental oversight placed few obstacles in the way of developers and industrial firms. The construction of an extensive transportation network, composed of railroads, electric streetcars, and even interurban electric railways, facilitated this dispersal of manufacturing activity and the growth of related industrial communities well before automobiles and trucks affected the metropolitan geography.

This geographical process of city building, or geographical industrialization, was more than simply centrifugal expansion from the city and new development in the suburbs. It involved the creation of a complex division of labor with multiple linkages among regional firms. Four industries formed localized production systems that, through their internal interrelationships, tied the spatially disparate regional parts into a metropolitan whole. The iron and steel, glass, and railroad equipment industries evolved into diverse complexes of large mass producers, specialized operators, subcontractors, suppliers, and servicers. This metropolitan growth of the Pittsburgh region eventually encompassed the extant Connellsville coke district, a fourth localized production system, which had initially been located beyond the metropolitan area.

Pittsburgh between 1850 and 1870

By the beginning of World War I, metropolitan Pittsburgh extended thirty to forty miles beyond downtown, a process that took nearly five decades to unfold. Before this metropolitan expansion took place, new growth both increased density within already developed areas and pushed the urban margin outward a few miles. In 1850 the City of Pittsburgh was a small commercial city of around 46,000 inhabitants situated on the point of land formed by the convergence of the Monongahela and Allegheny Rivers. Hilly topography and the three rivers (the Ohio River is generally viewed as commencing at the confluence of the other two) fragmented urban development such that growth also occurred in politically distinct communities on the floodplains directly across the Monongahela and Allegheny from Pittsburgh. Vigorous iron, glass, and textile industries hugged the rivers' edges of these communities, where flat land and transportation were available. To the north across the Allegheny River, Allegheny City experienced a quickening of manufacturing after 1834, when the Pennsylvania Main Line Canal was opened within its municipal limits. In 1850 this sister city had a population of 21,262, nearly half that of the City of Pittsburgh. To the south across the Monongahela, three much smaller towns were showing signs of industrial growth, especially with coal being mined in the steep hillside that abutted their southern edges. Together South Pittsburgh, East Birmingham, and Birmingham contained only 7,239 residents. The entire urban complex formed a small compact area of approximately 75,000 residents.[6] The only urban industrial development beyond, but near, the Pittsburgh area occurred in four tiny communities along the Pennsylvania Main Line and Beaver Canals.[7]

Fueled by rapid growth in iron, glass, oil refining, and planing mill activities, the urban area's population doubled to more than 170,000 during the next two decades. Although larger plants and some new product lines, such as blast furnaces for making pig iron, were added, this round of investment did not transform the basic organization of production. Most rolling mills, foundries, crucible steel works, and glass factories relied on longstanding craft practices, produced modest outputs, and expanded production more by addition than by technological innovation.[8] New plants were located farther away from Pittsburgh's central business district along the riverfronts. Andrew Carnegie and other iron masters built their works eastward up the south bank of the Allegheny River, where the Pennsylvania Railroad shared the flat floodplain. In partnership with several associates, Carnegie controlled the Iron City Forge and Cyclops Iron Company (together better known as the Union Mills, which produced bar iron, railroad car forgings, plate, and structural iron shapes), the Keystone Bridge Company (customer of the Union Mills), and the huge new Lucy (iron blast) Furnace.[9] Across the Allegheny and curving down the Ohio River a short distance, Allegheny City nurtured a sizable and diverse manufacturing base.[10] Iron and glass works multiplied along the south bank of the Monongahela River in three communities that became the city's South Side in 1872. Here, along the Monongahela, Benjamin F. Jones rebuilt the American Iron Works (later Jones & Laughlin Steel) into an integrated iron mill during the 1850s and 1860s; the mill included blast furnaces across the river in Hazelwood, then an emerging suburb contiguous to Pittsburgh.[11]

Although Jones's move to integration foreshadowed events that would transform the metropolitan geography, the spatial expansion of industry before the 1870s was for the most part incremental and contiguous. Manufacturers found available land with access to water and rail transportation at the fringes of the urbanized area on the river floodplains. Plants were getting bigger but, significantly, land was not available in the topographically constrained and intensely congested commercial city without expensive acquisition and demolition costs. Even with this initial round of industrial expansion into the suburban edges, the urbanized area remained tightly clustered within three to four miles of downtown.

The adoption of steam railroads and horsecars in the 1850s and 1860s, respectively, also spurred some residential suburbanization, but there was only a little industrial development beyond the dense core of urban settlement. A few modest foundries, rolling mills, and glassworks operated at noncontiguous river sites as much as ten miles from Pittsburgh. Populations of the accompanying towns were 3,500 or less. Scattered more widely around southwestern Pennsylvania were numerous small artisanal establishments and a few firms with more than twenty employees that produced pig iron, rolled iron bars and sheet, made glasswares, or baked coal into coke.[12] Often trading in the Pittsburgh market, these operations were too far afield to be part of the urban area, however broadly conceived.

Technological Change and Industrial Suburbs

By the early twentieth century, capital-intensive mass-production plants, often under corporately organized firms, replaced smaller traditional partnerships and craft factories as the dominant manufacturing form in the Pittsburgh area. This transformation of the region's industry, and ultimately of its urban form, began in the 1870s. Technological change, seen most clearly in the success of Bessemer steel production and unprecedented blast furnace outputs, introduced mass production to Pittsburgh's major industries, led to an enormous growth of manufacturing, and heightened the demand for large industrial sites. Integration of several manufacturing functions at one site, especially in the steel industry, exacerbated the pressure for land to the point that companies and real estate developers were speculatively buying up flat sites by the 1890s in anticipation of continued industrial suburbanization. Even though the search for suitable sites amid the region's rugged topography pushed industrial development well beyond the urban core, manufacturers generally concentrated along transportation corridors of river and railroad service in a manner similar to the industrial patterns of other cities. The simultaneous development of clay, coal, and natural gas resources both hastened the spread of industry and accentuated the importance of transportation links between resources and industrial plants. The pace and pattern of this industrial metropolitanization was uneven, as different industries and resources experienced these changes at different times.

The railroad dispersed industry around Pittsburgh as much through its role as a market for iron and steel as through its being a means of enhanced accessibility. The burgeoning demand for rails, cars, equipment, and bridges after 1850 dramatically expanded the market for iron and eventually set off a search for a way to produce large quantities of inexpensive steel, especially to replace notoriously unreliable iron rails. In the 1850s and 1860s Pittsburgh iron masters such as Benjamin Jones enlarged their

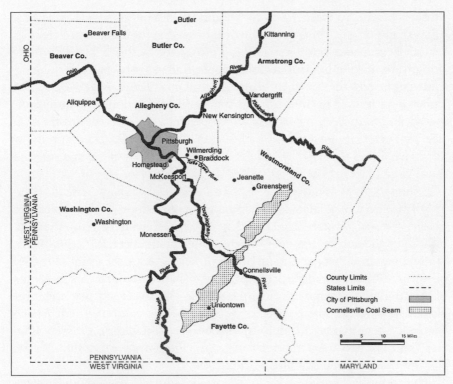

FIGURE 7.1. The Pittsburgh Metropolitan Area.

ironworks so as to participate in this market, but, with his close ties to the Pennsylvania Railroad, Andrew Carnegie entered the national competition to perfect the Bessemer steel process and manufacture steel rails. For the construction of an integrated Bessemer steel rail mill Carnegie chose in 1872 a greenfield site located nine miles from Pittsburgh at the junction of two railroads (one of them the Pennsylvania Railroad, which Carnegie anticipated would be a major customer) with the Monongahela River. The renowned Bessemer engineer Alexander L. Holley designed the works to fit the railroad so as to achieve a high-volume throughput, and within months of beginning production in 1875 the Edgar Thomson Works at Braddock became the model for the industry, pushing competitors to redesign their operations (see Figure 7.1).[13]

Although specialized and flexible batch production operations always found niches in the regional metals industry, mass production works proliferated under the aggressive investing practices and careful coordination of Pittsburgh capitalists and financial institutions. Large plants, whether those of Carnegie Steel, Jones & Laughlin, Pittsburgh Steel, or several other corporations, employed from 500 to several thousand employees. A few steel companies put together giant integrated facilities that did everything from making pig iron to rolling semifinished steel products. These same companies often owned or invested in coal mines, coke works, and iron ore reserves. This new form and scale of production frequently separated the location of ownership and management from the production site. Financiers in downtown Pittsburgh and other cities, especially New York, marshaled the huge capital resources necessary to build and operate these large plants at sites throughout the metropolitan region.

Although massive integrated iron and steel works characterized Pittsburgh in the national imagination, the metropolitan industrial complex was in fact vast and diverse. It included both a broad array of metalworking industries and substantial operations in aluminum, electrical equipment, glass, coke, machinery, and railroad equipment.[14] In most of these industries capital-intensive mass-production plants also dominated. The plants of Westinghouse and the Aluminum Company of America rivaled the integrated steel works in size, but many other firms, among them Mesta Machine (steel-making machinery), Dravo (boat building), Pressed Steel Car (railroad cars), Armstrong Cork, and Pittsburgh Equitable Meter, employed a thousand or more workers.

The adoption of mass production, integrated works, and modern management led many Pittsburgh manufacturers to search for large sites with railroad and often river accessibility. Some firms, such as Carnegie Steel, Jones & Laughlin Steel, and George Westinghouse's Air Brake and Electric companies, developed new locations beyond the city in order to modernize and expand their operations. They purchased land (usually floodplains along a waterway), designed plants to modern production standards, and sometimes built adjacent towns to house workers. In 1890, for example, Westinghouse moved his railroad air brake factory from Pittsburgh to a site twelve miles away in the Turtle Creek Valley, a tributary of the Monongahela River through which the main line of the Pennsylvania Railroad ran (Figure 7.1). Next to his modern factory and company headquarters he built the town of Wilmerding for the workers. Four years later he moved his electrical manufacturing company from Pittsburgh into an enormous factory at a site a few miles downstream from the air brake plant.[15] In 1904 he constructed a foundry to supply castings for giant generators and an accompanying town (Trafford) immediately upstream from Wilmerding. With more than 20,000 workers in these three plants in 1919, the Turtle Creek Valley became known locally as the electric valley.[16]

In the half-century since its formation before the Civil War, the Jones & Laughlin Steel Company (J & L) expanded, modernized, and reorganized its works on both sides of the Monongahela River in the City of Pittsburgh, but topography and city neighborhoods constrained further growth at these sites. Additional expansion and the construction of new mills for entry into markets not served by its Pittsburgh Works required the development of a new site. In 1905 J & L purchased a farm and dilapidated amusement park on the floodplain of the Ohio River several miles below Pittsburgh. There the company erected blast furnaces, open hearth furnaces, and, by World War I, tin, rod, wire, nail, and blooming mills. A subsidiary company laid out the town, built the houses, and sold them to J & L employees. By 1920 more than 15,000 people resided in the new communities of Woodlawn and Aliquippa adjacent to the Aliquippa Works.[17]

Investors attracted others firms to new communities begun as speculative industrial real estate ventures. For example, in 1894 a group of Pittsburgh investors purchased a farm on a floodplain nestled in a broad curve of the Monongahela River thirty miles south of the city. Although they correctly anticipated the expansion of industry up the river, they resorted to financial inducements to get their development underway. Learning through personal connections that William Donner was considering moving his tin plate mill from Indiana to the Pittsburgh district, the group offered him twenty acres for free and a $10,000 cash bonus. Donner's new plant opened in 1898 and by 1903 it was the largest tin plate mill in the nation. Meanwhile, the investment group formed the East Side Land Company to lay out and sell land for an adjacent community. By World War I their venture had become the city of Monessen (Essen on the Mon), home, in addition to the tin plate mill, to a foundry and machine firm, steel hoop plant, rod and wire mill, tube works, and a fully integrated steel works employing 3,500 workers in 1916 (Figure 7.1). With more than 18,000 inhabitants, Monessen surpassed New Kensington and the county seat of Greensburg as the largest city in Westmoreland County.[18]

In some cases the interests of the relocating firms and the real estate developers were closely allied. Originally nurtured in Pittsburgh's "strip district" along the Allegheny River adjacent to downtown (as Westinghouse and many other firms were), the Pittsburgh Reduction Company, later renamed the Aluminum Company of America (ALCOA), selected New Kensington in 1891 for the site of its modern aluminum plant (Figure 7.1). Located nearly twenty miles up the Allegheny River in Westmoreland County, New Kensington was the speculative real estate development of Pittsburgh-based Burrell Improvement Company. The new location provided access to natural gas and coal for the huge amounts of power required in aluminum manufacturing. An additional incentive for the struggling aluminum company came from Burrell Improvement, which offered four free acres of level land and a $10,000 cash bonus. Instrumental in this deal were the Mellons, who were investors in both the Pittsburgh Reduction Company and the real estate firm. Moreover, they advanced a loan to underwrite the costs of the move. In time New Kensington became the center of a small cluster of industrial towns that were home to a specialty alloy steel works, tin plate mill, spring company, window glass factory, and several smaller firms. By 1919 ALCOA alone employed 2,400 workers, and the New Kensington area had grown to approximately 18,000 residents.[19]

While the move to mass production was the primary impetus for greenfield development at the turn of the century, management sometimes had social and political considerations as well. Pittsburgh had long been a center for the craft union movement and experienced years of contentious labor relations. Some owners wished to remove their plants and workers from this cauldron of labor politics by relocating to isolated communities where they could exercise more control. After a bitter strike in 1893, for example, the Apollo Iron and Steel Company relocated to a farm site almost directly across the Kiskiminetas River from its cramped Apollo works. Here the Pittsburgh-owned company built a modern works that took advantage of the greater available space, but management also wanted to stabilize the labor force through the construction of a model industrial town, designed by Frederick Law Olmsted's well-known architectural firm. With an attractive plan and modern infrastructure, company president George G. McMurtry hoped to encourage workers to become homeowners and turn their backs on the unions that had beleaguered the original works across the river. By 1920 more than 12,000 people resided in the industrial communities of the Vandergrift area.[20]

Similarly, B. F. Jones Jr. and William Larimer Jones, founding family members, partners, and chief officers of J & L Steel, were influenced by Pittsburgh's social gospel reform movement and saw the advantage of welfare capitalism then being instituted by their primary competitor, the United States Steel Corporation, among others. They endeavored to establish both decent housing and a decent community for workers and their families by controlling the city-building process and dominating local institutions. Characteristically, however, such company domination of industrial communities became a two-edged sword; company policies often had both benign and pernicious effects. The police force of the Aliquippa

Works, for example, worked closely with local police to enforce the company's vision of proper behavior and keep out undesirables. The line between maintaining civil order and impinging on individual freedoms was blurred, and Aliquippa soon gained a reputation for being hostile to unions and union organizers.[21]

Carnegie's famous Homestead Works, the site of the violent 1892 Homestead Strike, exemplifies the political advantage of some suburban locations. A group of Carnegie's competitors originally built the mill in the early 1880s at a greenfield site along the Monongahela River in view of Carnegie's Edgar Thomson Works on the other shore, which they wished to challenge. However, poor management and a recession provided Carnegie the opportunity to buy the mill. Homestead was a short eight-mile train ride from the city. Control of the railroad station provided control over ingress and egress to the town, although management, the union, or the state militia could exercise this power, as strike events later proved. The steel works, however, came to be located in adjacent Munhall, a municipality carved out by the company in 1901, where it built housing for management and controlled municipal government to its advantage. The workers mostly lived in Homestead without municipal control over either the plant or its potential tax revenues. Through the years both Munhall and Homestead depended on the company's largesse for many infrastructural improvements and annual maintenance.[22]

Natural Resources and Metropolitan Development

Although the region's natural resources—timber, coal, iron ore, and clay—had been used since the beginning of the nineteenth century, the industrial transformation after 1870 led to the massive exploitation of coal beds, clay deposits, and natural gas reserves. Small entrepreneurs, often indigenous to the resource's locality, leaped at the opportunities to supply the region's growing industries, especially with coal and coke. Large capital interests viewed these resources as one more factor of production to be controlled, exploited, or held for speculation, however. Pittsburgh investors such as Judge Thomas Mellon amassed hundreds of acres of coal lands throughout the metropolitan region, and hired operators to mine them or left them undeveloped in order to realize profits from rising land values later.[23] Industrial firms such as Carnegie Steel and J & L sometimes developed their own "captive" mines, coke works, gas wells, or quarries and integrated them closely with their manufacturing operations. Whether owned by independent operators, speculators, or industrial firms, the intensive development of resource sites contributed to the sprawling character of the metropolitan region.

Clay, natural gas, and coal attracted manufacturers to sites widely scattered about the region. Clay deposits supported a sizeable number of refractory brick manufacturers. Blast furnaces, various iron and steel furnaces, glass furnaces, beehive ovens for making coke, steam engines, and other high-heat apparatus all needed fire brick linings. By 1919 approximately three dozen plants employed a few thousand workers. Ceramic firms also capitalized on regional clay resources, building plants around the region to make pipe, tile, electrical fixtures, and tableware.[24]

Natural gas had its greatest impact on the glass industry. In 1880 Pittsburgh's nationally important glass industry remained concentrated on the city's South Side in fifty family-owned, small-batch-oriented factories. The city contained both a large number of skilled glass workers and numerous equipment and supply firms. Beginning in the 1880s, the advantages of natural gas as a fuel and the adoption of mass production technologies led manufacturers to build large factories in the gas fields of southwestern Pennsylvania as well as beyond the metropolitan region. In 1888, for example, prominent Pittsburgh glass manufacturers H. Sellers McKee and James Chambers purchased farmland twenty miles east of the city, where they could spread their new integrated window glass factory over thirty-five acres to accommodate continuous throughput production, enjoy access to the Pennsylvania Railroad's mainline, and obtain a steady fuel supply from nearby gas deposits. These entrepreneurs developed the property into the city of Jeannette, which, with the addition of other firms, grew into an important center of glass manufacturing and a community of more than 10,000 in 1920 (Figure 7.1). Many other sites for window, plate, and glassware manufacturing emerged around the metropolitan area. By one count, only fifteen plants remained in the city in 1902, while sixty-three plants operated in the metropolitan region.[25] Pittsburgh's last window glass factories closed in 1907, while the region retained 25 percent of the nation's capacity.[26]

Like clay and gas, coal spawned development at the site of the resource. In the first half of the century mining occurred in Pittsburgh and the immediate environs as well as at sites scattered around the region. Most mines were small and not necessarily linked to Pittsburgh markets and capital. The situation changed dramatically after 1860. The rapid expansion of Pittsburgh's iron industry in the 1860s overwhelmed the pig iron output of the rural, charcoal-fueled blast furnaces that traditionally supplied the city's rolling mills. In order to operate their own blast furnaces, city iron masters had to find a fuel other than wood charcoal that would be chemically appropriate and capable of transport to the city. Coke was already being manufactured near mines around the region, including some in Pittsburgh, for use primarily in foundries and forges because it had not proved suitable for blast furnaces. In 1860 Pittsburgh iron masters discovered the superior qualities of coke made from coal found in a narrow forty-mile-long seam running parallel to the western front of the Allegheny Mountains thirty to forty miles southeast of the city. This area became known as the Connellsville coke district (Figure 7.1). Connellsville-area investors built 500 to 750 beehive ovens during the 1860s as the industry took its first tentative steps.[27]

Railroad construction into the Connellsville district and the emergence of the American steel industry, especially in Pittsburgh, transformed this rural area during the next three decades. In the 1870s the construction of huge blast furnaces in Pittsburgh to feed pig iron to the voracious Bessemer converters set off both a coal rush and the new coke industry.[28] Investors snapped up these special coal lands, opened dozens of mines, and erected thousands of beehive ovens, which baked the coal into coke and released the unwanted gases into the air. More than 20,000 ovens and ninety-four mines produced half of the nation's total coke output in 1900.[29]

Pittsburgh capitalists joined with indigenous Connellsville district investors to undertake and profit from the development, although the importance of Connellsville coke to American steel producers attracted capital from outside the region as well. H. C. Frick, for example, was living and working in the district when he initially entered the industry. With financial backing from Pittsburgh's Thomas Mellon and other city investors in the 1870s, he emerged as a powerful force in the coke district. A partnership with Andrew Carnegie, formed in 1881, after Frick had moved his residence to Pittsburgh, provided the capital to propel the H. C. Frick Coke Company to the dominant position. The leading figures of these two key regional industries were linked in an uneasy arrangement that led to Frick's appointment as president of Carnegie's steel empire, a position he held until the late 1890s, when they bitterly parted ways.[30]

When the Connellsville coal resources began to show some signs of exhaustion around the turn of the century, coal producers drifted a few miles west into the "Klondike" or Lower Connellsville area. In the peak year of 1910 operators from the two areas worked more than 39,000 ovens and shipped nearly 19 million tons of coke. Later coke operators crossed the Monongahela River to open mines with coal that could be made into suitable coke with the byproduct retort technology that replaced beehive ovens.[31]

At the time of its initial exploitation around 1870, nearly thirty miles separated the Connellsville district from the Pittsburgh iron market. In subsequent decades the development of a dense web of railroads, close capital and labor linkages, and steel works in the middle Monongahela River Valley drew the two areas tightly together both spatially and functionally. The Frick-Carnegie partnership symbolized this convergence. The Monongahela River was especially attractive to primary iron and steel producers because it knifed through the heart of the region's largest coal reserves and was close to the Connellsville district. By World War I iron and steel works occupied noncontiguous floodplain sites of this meandering river for almost thirty miles south of Pittsburgh's downtown. Monessen, for example, was as little as ten miles from Connellsville district coke plants.

The massive development of coal and coke after 1870 spawned manufacturing opportunities in nearby towns and cities of Westmoreland and Fayette Counties. The small cities of Connellsville, Uniontown, and Greensburg, which served as centers of administration, capital, and supply for the coke district, thrived on this new industry. Local foundries and machine shops produced and repaired mining cars and equipment, and a few even manufactured railroad cars for local coke firms. Railroads built spur lines throughout the coke district to service the local mines and coking plants. This rail network linked the coke area into the Pittsburgh market and after 1880 supported the development of local metalworking firms making specialty alloy steels, steel-making machinery, railroad equipment, pipe, valves, boilers, and plumbing supplies.[32]

The Metropolitan Geography

Entrepreneurs in the other metropolitan counties of Beaver, Butler, and Washington similarly found market niches for their products. Together with Pittsburgh's manufacturers and those who left the city for, or originally built in, outlying industrial suburbs and satellites, they formed the nation's sixth-largest urban manufacturing district in 1920 and fifth-largest metropolitan area. The region's rugged hill and valley terrain inhibited simple, incremental spatial expansion at the city's periphery and forced firms or investing groups to find suitable sites that were often not contiguous with the advancing edge of urban development. While Pittsburgh formed a dense urban core for the region, industrial towns and residential suburbs stretched linearly away from this core along the major mean-

TABLE 7.1. Population Distribution of Metropolitan Pittsburgh, 1900–1920

Type of community	% of metropolitan population	
	1900	1920
City of Pittsburgh	58	45
Residential suburbs in Allegheny County	5	8
Industrial suburbs in Allegheny County	20	23
Industrial suburbs contiguous with Allegheny County	9	13
Satellite cities	8	11
Total metropolitan population[a]	776,458	1,312,017

Source: *Twelfth Census of the United States, 1900, Population*, I (Washington, D.C., 1902), 329–49; *Fourteenth Census of the United States, 1920, Population*, I (Washington, D.C., 1922), 586–601.
[a]Excluding coal-patch communities.

dering river valleys and the main railroads that snaked through smaller tributary stream valleys. A series of satellite cities in the surrounding counties and the industrial settings at the clay and coal resources areas completed the metropolitan pattern.

Between 1870, when the centrifugal movement of industries accelerated, and 1900, most industrial development occurred in Allegheny County. At the beginning of this period Pittsburgh had more than 90 percent of Allegheny County's urban population and, by the most liberal definition of "metropolitan," 80 percent of the regional urban population.[33] There were only six very small industrial communities outside the city in Allegheny County. By the turn of the century there were twenty-one industrial towns and boroughs outside Pittsburgh in the county. Another dozen or so communities in the county were residential suburbs, but they contained approximately one-quarter of the population of the industrial suburbs. Together these industrial and residential suburbs diminished the city's proportion of the regional urban population to 58 percent in 1900 (see Table 7.1).[34]

Industrial suburbanization burgeoned in the next two decades, with an expansion of more than 300,000 in population. The number of industrial towns and cities with more than 5,000 inhabitants nearly doubled from twenty-three to forty-five in 1920 (see Table 7.2). Seven of these cities were larger than 20,000. Large new steel works opened at six towns in the Monongahela River Valley, and four new ones were started in the Ohio River Valley. New industrial towns with manufacturing plants established in the 1890s also grew rapidly during the early decades of the twentieth century. The aluminum and glass towns of the Allegheny River Valley and

TABLE 7.2. Town Size Distribution of Metropolitan
Pittsburgh, 1900–1920

Population size	Number of towns	
	1900	1920
50,000+	1	1
20,000–49,999	2	7
10,000–19,999	4	16
5,000–9,999	16	21
Total number	23	45

Source: Twelfth Census of the United States, 1900, Population, I
(Washington, D.C., 1902), 329–49; *Fourteenth Census of the
United States, 1920, Population,* I (Washington, D.C., 1922),
586–601.

the Westinghouse towns of the Turtle Creek Valley both exemplified this
latter pattern. Overall, the City of Pittsburgh's share of the metropolitan
population declined to 45 percent, while the proportion of population in
industrial towns outside Allegheny County rose from 17 percent to 24
percent. Although residential suburbs increased in number and size, they
still were unable to match the growth of the metropolitan industrial communities. Dormitory suburbs made up approximately 8 percent of the
metropolitan population. The complex topography notwithstanding, there
was a discernible distance-decay pattern of development spreading outward
from the city. The City of Pittsburgh contained 45 percent of the metropolitan population, adjacent industrial suburbs in Allegheny County added
23 percent, industrial towns contiguous to but outside Allegheny County
had 13 percent, and the more distant satellite cities and towns had another
11 percent.[35]

The spatially extensive and complexly patterned geography of the Pittsburgh industrial district led to confusion among the Pittsburgh Survey
fieldworkers, local politicians, and journalists as to what actually constituted the metropolitan region.[36] Were the many communities not contiguous with Pittsburgh and its adjacent urbanized areas part of it? Survey
editor Paul U. Kellogg argued that "today, for most purposes, a city is a
rapid transit proposition."[37] As a progressive reformer, he was concerned
with political consolidation for the sake of more efficient and effective
municipal governance. His spatially restrictive perspective, one that apparently emphasized residential suburbanization, ignored the multifarious
relationships that existed among the various parts of the industrial district.

Metropolitan Integration

What contemporaries and later historians often viewed simply as industrial decentralization in fact involved, in many instances, the creation or expansion of localized production systems. In Pittsburgh the rapid growth of iron and steel, glass, railroad equipment, and coke industries under corporate capitalism resulted in both integrated mass production plants and an increasing division of labor among firms, involving specialization, subcontracting, and services. As capital deepened and interdependence among firms grew, the number of participants increased, external economies from agglomeration accrued, and localized production systems formed around these industries. The flow of goods and services, capital, labor, and professional personnel through the dense transportation and communication networks among the many firms of these localized production systems tied the disparate sites of production and their communities into a metropolitan region without regard to spatial contiguity.[38]

The iron and steel industry was by far the largest localized production system of the Pittsburgh region. The seemingly self-contained character of the large integrated steel works masked the persistence of older specialized batch producers, the division of labor, and the extent of linkages between firms. While the huge scale and dramatic technological spectacle of integrated steel mills at Braddock, Homestead, Duquesne, the South Side, Monessen, and Aliquippa drew popular and later scholarly attention, the vast array of iron and steel activities emerged in relative obscurity. Independent mills producing wrought-iron products or mills that had switched to open-hearth steel production outnumbered and collectively out-produced the fewer giant integrated operations.[39] Hundreds of other firms supplied or serviced the basic iron and steel mills or used the semi-finished products in their own operations. Foundries produced custom castings and steel-making machinery. Machine shops made and repaired engines and custom-tooled machinery and parts. Other firms manufactured pipes, tubes, wire, fencing, tanks, barrels, plumbing supplies, hardware, barges and boats, and countless other items. Model and pattern makers, engineering firms, steel product distributors, scrap dealers, steel construction and fabrication firms, haulers, and many others were scattered throughout the region. The density of metalworking and metal-serving firms resulted in the development of secondary specialties, such as bridge building and engineering, which gained national renown. Corporate administrative offices, investment banks, and legal services took root in downtown Pittsburgh. Membership in investment alliances and boards of directors overlapped and frequently changed in complex patterns.[40] Even other local industries, such as refractory bricks and the railroads, were tied directly to the iron and steel complex, supplying goods and services or buying steel products.[41] The interrelationships of the numerous firms involved in this complex made up a metropolitan-wide, localized production system just as surely as did the better-known smaller-scale garment districts of New York and Los Angeles.

The glass, railroad equipment, and coal and coke industries also displayed the scale and division of labor of localized production systems. Railroads were so pervasive throughout the region that one may easily overlook the substantial metropolitan railroad equipment and repair industry. The extensive rail network involved the trunk lines of six major railroads, including the Pennsylvania and the Baltimore and Ohio, numerous feeder roads, especially from coal mines, short lines between towns, and more than two dozen industrial switching railroads connecting separate plants of an individual firm or manufacturing plants with trunk lines. A dozen or more repair shops and switching yards, some quite large, operated around the region, employing 4,500 workers in 1902.[42]

This concentration of railroads, coal mines that used railroad technology, and metal manufacturers led to a sizeable railroad equipment manufacturing sector. Besides the rails produced by iron and steel mills, companies in the region turned out locomotives, steel cars, air brakes, couplers, switches, signals, wheels, axles, frogs, and many other items. Both locomotive manufacturers and car makers centered distinct clusters of suppliers in at least four locations in and around the city, each of which employed thousand workers. The H. K. Porter Company, for example, built narrow-gauge locomotives for switching, mine hauling, logging, and other uses. Producing heavily for regional customers, H. K. Porter remained in the city, along the Allegheny River, where rolling mills, wheel, spring, coupler, and other equipment makers (including Westinghouse's original companies) were concentrated. While approximately 7,100 workers manufactured railroad equipment in this cluster in 1902, the Pittsburgh Locomotive Works was situated on the city's North Side amid a similar cluster of manufacturers employing nearly 10,000 workers.[43]

In 1919 more than four dozen glass manufacturers of various products operated within the region. Although many of Pittsburgh's glass firms had left for other metropolitan locations (or left the region altogether), the intricate interrelations among the various components of the industry remained intact. Historian Richard O'Connor even found the city important to the window glass industry, which had entirely moved out. "Although window glass manufacturers abandoned the city proper, they found the environments of outlying areas hospitable. By locating their large, technologically advanced, firmly capitalized plants within the orbit of Pittsburgh capital, labor, and suppliers, they extended the city's sphere of influence without formerly redrawing its borders."[44]

Iron and steel as a localized production system encompassed the entire metropolitan area. Individual firms were connected physically by railroad, river, telegraph, telephone, and postal services. Capital also linked firms through emerging corporate organizations with interlocking boards, private investment banks (e.g., T. Mellon and Sons), securities markets (e.g., the Pittsburgh Stock Exchange), and social institutions (e.g., prestigious business clubs such as the Duquesne Club and golf clubs).[45] These communication networks allowed administrative offices and investors to keep a close watch over their metropolitan plants. Owners and managers located at their plant sites also enjoyed easy access to downtown headquarters, business services, and entertainment opportunities. In 1922, in addition to electric trolleys and interurbans, 368 commuter trains ran daily between downtown Pittsburgh and the several satellite cities within a forty-mile radius, making them less than an hour's ride from the city.[46] Even the network of trolleys tying the coal-patch towns of the Connellsville coke district together was linked to the broader metropolitan region through short direct routes to both McKeesport and Trafford, industrial towns only fifteen miles from Pittsburgh.[47]

Labor markets also manifested the systemic nature of regional industry. Employers and employees alike benefited from the concentration of firms in specific industries, employers from the reservoir of experienced workers, employees from multiple job opportunities. The effects of these links were not always beneficial, however. Steel workers feared blacklisting by employers, who commonly used the tactic to intimidate labor. Unions understood that part of their leverage lay in their ability to affect production at proximate and interconnected facilities. The 1892 Homestead Strike failed, in part, because Carnegie was able to keep his other local plants at Braddock, Duquesne, and Pittsburgh operating.[48] In 1916 Westinghouse Electric's striking machinists first attempted to close down other Westinghouse plants in the Turtle Creek Valley and elsewhere in the region, with some success. When they then marched down the valley, hoping to paralyze Carnegie's steel works at Braddock, violence broke out but Braddock stayed on line.[49] Fighting the United Mine Workers at their own (captive) mines, U.S. Steel justifiably feared that their miners would traverse the dozen or so miles to their Clairton steel works to disrupt its operation.[50]

In his seminal 1933 study *The Metropolitan Community*, R. D. McKenzie observed that the modern metropolitan region encompassed an unprecedented amount of territory and obtained "its unity through territorial differentiation of specialized functions," the speed and intensity of communication, and the interdependence of these functions.[51] He believed that this far-flung metropolitan community was the result of the integration of automobile and truck transportation and electronic communication into everyday life after World War I. "Under present conditions of local transportation," he wrote, "the population residing within a radius of 30 to 40 miles of a large city may be considered as within direct contact with its institutions and services and therefore in a sense might well be included as inhabitants of the city's metropolitan district."[52]

It is clear, however, that both the centrifugal spread of industry and the development of localized production systems between 1870 and 1920 created a sprawling and highly interdependent metropolitan region in parts of several counties of southwestern Pennsylvania before the widespread adoption of motor vehicles. The Pittsburgh district, as contemporaries called it, presented a complex urban landscape, with a large dominant central city surrounded by proximate residential suburbs, mill towns, small satellite cities, and hundreds of mining patch towns. Although ambiguities of both census definitions and the realities of land use defy accurate measurement, residential suburbanization certainly took a backseat to the centrifugal impetus of industry in forming Pittsburgh's metropolitan geography.[53] While commuters used steam railroads and horsecars to push the city's margins into the surrounding countryside, the region's financiers, speculators, and industrialists aggressively developed infrastructure, natural resources, and hospitable floodplain sites well beyond the city's limits. Between 1870 and 1900 their actions erected the outlines of a vast industrial metropolitan region, which would be filled in during the next twenty years of sustained growth. The city simply could not contain the requirements of the new industrial order of mass production, corporate organization, rapid communication and movement, and labor control. The region's distribution of natural resources and difficult topography certainly accentuated the spatial sprawl of Pittsburgh's industrial development. But, as can be seen in other cities with simpler, more level environments, the sheer magnitude of industrial growth and profitability of speculative real estate development would still have created a complex metropolitan district, one that cannot be adequately described or explained by traditional views of suburbanization.

8

The Suburbanization of
Manufacturing in Toronto,
1881-1951

GUNTER GAD

n 1881 Toronto was a medium-sized city with a population of 86,000 and a very generous political territory of twenty-three square kilometers. Approximately 13,200 workers were employed in some 930 manufacturing establishments within the boundaries of the city of Toronto, with workshops and factories spread throughout most of this area.[1] Clothing factories and printing workshops were found above and between the retail, wholesale, and office establishments of the central business district. Brass foundries, flour mills, furniture and piano factories, and carriage makers clustered east and west of this district. Farther afield, mostly along the railway tracks fanning out from the city's center and out to the edge of the built-up area, there were stove foundries, machinery works, breweries, a major whiskey distillery, and meatpacking plants, the last of which had also attracted tanneries and soap manufacturers. Small clusters of planing mills, sash and door factories, piano factories, and other establishments, largely in the wood processing industries, were scattered across the city. Many of the more peripheral factories had recently moved from more central locations; others were new arrivals from small towns in southern Ontario; and still others, like the breweries and the whiskey distillery, were established between the 1840s and 1880s, well in advance of other

businesses in the built-up area.[2] This early spatial differentiation of manufacturing gradually expanded over a much larger area in the following
decades. Between 1881 and 1951 manufacturing employment in Toronto
grew from 13,200 to nearly 200,000, and the semicircle over which factories were spread increased in radius from 3.5 to 12 kilometers (see Figures
8.1 and 8.2).

A description of suburbanization proper, that is, of the spread of population and manufacturing activity beyond the boundaries of the "central
city," must deal not only with the spatial dynamics of manufacturing but
also with the instability of political boundaries. In Toronto, earlier phases
of suburban development were incorporated into the city through two
waves of annexation, the first between 1883 and 1893 and the second
between 1905 and 1914 (Figure 8.1). These annexations resulted in a fairly
large political territory of 79.2 square kilometers. Between 1914 and 1951
the boundaries of the city of Toronto remained the same, while a new gen-

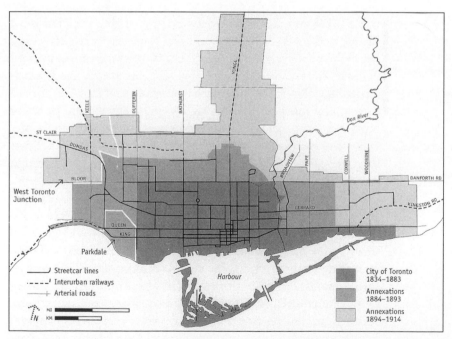

FIGURE 8.1. City of Toronto, Early Suburbs and Annexations to 1914.
Source: R. C. Harris, F. A. Gaby, and E. L. Cousins, *Report of the Civic Transportation
Committee on Railway Entrances and Public Transit for the City of Toronto*, vol. 2: *Plans*
(Toronto, 1915). Plans on annexation and on street and interurban trackage.

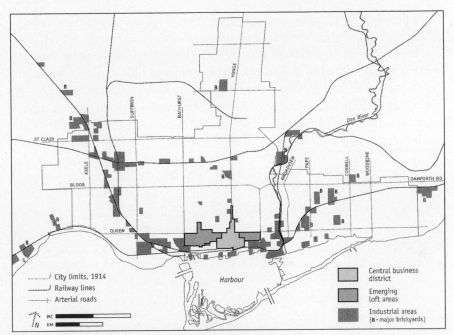

FIGURE 8.2. Industrial Areas in the City of Toronto, 1914.
Sources: R. C. Harris, F. A. Gaby, and E. L. Cousins, *Report of the Civic Transportation Committee on Railway Entrances and Public Transit for the City of Toronto*, vol. 2: *Plans* (Toronto, 1915). Plan on development and property; *City of Toronto Directory*, 1915.

eration of suburbs developed. Even in this new phase of suburbanization, however, Toronto was relatively large in terms of territory and population in relation to the suburbs. The degree of suburbanization of manufacturing had been very small between 1881 and 1914 and then had increased gradually to about 20 percent of manufacturing employment of the whole urban area by 1951.

The primary concern of this chapter is to explore the limited extent of the suburbanization of manufacturing employment in the Toronto area between 1881 and 1951. Since little has been written on this subject,[3] a substantial part of the task is to provide a basic but comprehensive account of the urban development process in the Toronto area, including its social, political, and transportation geographies. The structure of Toronto's manufacturing sector is also very important. It provides clues not only about the extent but also about the character of industrial suburbanization. What were the spatial dynamics of different sectors or industries? Did the differential suburbanization process, together with the changing social and political geography, lead to the formation of suburban "industrial districts" or, beyond that, to "industrial suburbs" (industrial agglomerations that were complemented by the residential districts of blue-collar workers)? A further concern involves the "spreading out" of manufacturing within fixed city boundaries and at the periphery without a corresponding rise of separate political jurisdictions. This "spreading out" cannot be described under the conventional terms "suburbanization" or "decentralization."[4]

Theory, Time, and Place

The theoretical context for an account of the industrial suburbanization process in Toronto is provided by a number of approaches. Conventional explanations focus on changes in transportation technology (the advent of cars and trucks), production technology (new power sources such as electricity), and the advent of single-story factory buildings[5]. Although these are insufficient explanations of the suburbanization of manufacturing, they raise questions that will be pursued in this chapter. The "transactional" and "industrial district" approaches developed by Allen Scott, and especially the notions of the persistence of small-scale labor-intensive industries in central locations and the decentralization of large-scale capital-intensive industries after 1900 will be deployed to probe the dispersal process in Toronto.[6] While Scott's approach considers the locational importance of different kinds of labor in the context of the industrial accumulation process, it does not explore fully the links between place of work and place of residence.[7] Also, capital and labor require land and infrastructure. Questions of labor control in urban areas and provision of urban land and infrastructure have been addressed in a variety of political economy approaches.

An early formulation was provided by Gordon Garland in a study of industrial suburbanization in Toronto during the 1880s and 1890s.[8] According to this approach, the suburban fringe was extended by politically dominant landowners, whose interests converged with those of the ascendant monopoly capitalists. An incorporated suburban municipality was to provide the basic infrastructure for both residential and industrial development. Long-term tax exemptions and a range of other municipal concessions and services were to enhance the profitability of industrial capital while simultaneously generating demand for workers' housing, thus enhancing the value of suburban land and the fortunes of landowners and subdividers. This approach considers which kind of production would be able to take advantage of suburban sites. Garland argued that petty-commodity or craft production would stay in the "externality-rich" central city, while vertically integrated factory production would be able to migrate outward. Nevertheless, vertically integrated production in capitalist factories would rely on some collectively provided infrastructure, especially railway facilities, water supply, and roads. In the established city, with its plurality of interests (petty-commodity producers, commercial businesses, already established residents), the demands by the new capitalists would be challenged, while in the new suburb the capitalists would be able to cooperate with the landowners and subdividers and shape favorable conditions for capital accumulation.

The approach of this chapter takes into account a large set of events and circumstances in a complex urban area, including the capital-accumulation process, with its implications for "linkages" (flows of materials and products) and labor requirements. But production and labor are embedded in broader economic, social, and political frameworks, including the city-building process. It is not taken as given that the city-building process was always under the control of industrial capital. I make no attempt to rewrite theory, but the approaches sketched here have shaped data collection and the interpretation of the assembled materials.

Time and place are extremely important factors in industrial suburbanization. Since this survey covers a large span of time, different sections require specific contextual remarks. Several characteristics of the Toronto area, however, are constant over the study period, and in these cases long-term dynamics are appropriately discussed at the outset. First, there is the physical setting. Although Toronto is a lakefront city, the harbor and other waterfront locations did not play a significant role in the location of manufacturing, contrary to popular belief.[9] There were no canals, and the rivers were unsuitable for navigation and yielded only small amounts of waterpower. More important, since the 1850s railways had fanned out from the city center across an increasingly wealthy agricultural hinterland and became major attractions for manufacturers, especially those using bulky inputs and shipping large quantities of outputs. Apart from shales and glacial deposits, used in the production of building materials, there were no mineral resources.

Of great importance for the relatively low degree of suburbanization was the mix of industries in Toronto (see Table 8.1).[10] Between 1881 and 1951 highly labor-intensive industries like clothing and printing, which were typically centrally located within major urban areas at this time, formed a strong component of manufacturing employment in Toronto. Within the many other sectors making up Toronto's industrial structure there were

TABLE 8.1. Composition of Manufacturing Employment in Toronto, 1880, 1911, 1930, and 1951

Sector	City of Toronto (%)			Greater Toronto (%)
	1880	1911	1930	1951
Food and beverages	7.5	9.2	16.0	13.0
Tobacco	3.1		0.5	0.1
Rubber		0.2	2.7	3.2
Leather	11.3	4.8	2.1	1.8
Textiles	1.8	2.2	3.1	2.9
Clothing	20.6	25.8	19.2	11.3
Wood and furniture	11.2	6.0	4.3	3.4
Paper	1.6	1.3	3.1	4.4
Printing and publishing	13.5	7.5	10.7	10.4
Iron and steel/machinery	13.2	18.5	13.5	15.9
Transportation equipment	1.7	3.1	0.3	4.1
Nonferrous metals			3.5	4.5
Electrical appliances		0.4	5.2	10.9
Nonmetallic minerals	1.8	2.7	3.2	1.7
Petroleum and coal			0.0	1.5
Chemicals	1.8	1.8	4.3	5.4
Other	10.9	16.5	8.3	5.6
All industries	100	100	100	100
Number of employees	12,708	54,000	95,000	195,000

Sources: 1880: G. Gad, "Location Patterns of Manufacturing: Toronto in the Early 1880s," *Urban History Review* 22 (1994): 117, based on 1881 Census of Canada, 1880–1881. 1911: E. Bloomfield, G. Bloomfield, and M. Vallieres, "Urban Industrial Development in Central Canada: Manufacturing Structure, 1911" in *Historical Atlas of Canada*, vol. 3, ed. D. Kerr and D. Holdsworth (Toronto, 1990), plate 13 and unpublished table, Historical Atlas of Canada files, Cartography Office, Department of Geography, University of Toronto. 1930: Canada, *The Manufacturing Industries of Canada, 1930* (Ottawa, 1932), appendix 1, 77–79. 1951: Canada, *General Review of the Manufacturing Industries of Canada, 1951* (Ottawa, 1954), 118–20.
Note: The sectoral structure for 1880, 1911, and 1930 has been arranged according to the classification of sectors and industries in the 1951 tables of *The Manufacturing Industries of Canada*. The sectoral arrangement for 1911 relies on a tabulation of occupational data from the 1911 Census. Since some occupations, e.g., "labourer," could not be allocated to particular sectors, the total employment of 54,000 listed here is considerably lower than the 65,274 total for all of "manufacturing" employment.

few large-scale enterprises occupying extensive areas. For instance, there was only one steel-making factory in the metal-products sector. (This steel mill, using electric-arc furnaces, operated for only a couple of years during World War I.) There were no large chemical complexes; the chemical sector consisted mainly of soap and paint factories. In the transportation equipment sector, carriage works gave way to auto assembly plants, but these never lasted long. Only the World War II airplane factories (and other war industries) reached a very large scale, with employment levels of 5,000 to 10,000 workers at one site.

Throughout this seventy-year period Toronto's economy was diverse, and manufacturing was only one of its major components. Although Toronto was Canada's second-largest manufacturing center, manufacturing as a whole never accounted for more than 35 percent of Toronto's labor force. Toronto's class structure was strongly influenced by the city's role as a commercial and financial center and as the seat of the provincial government of Ontario. It is generally agreed that Toronto remained largely a British city until after World War II.[11] The moderate influx of non-British immigrants did not lead to large-scale social segregation. The social mosaic in the central city was finely grained and the suburban ring was diverse in social composition, ranging from affluent, well-organized suburbs to patches of peripheral urban working-class settlements equipped with the most minimal urban services right into the early 1950s.[12]

Suburbanization of Manufacturing and the Growing City, 1880–1914

Between the early 1880s and the mid-1910s, manufacturing employment in Toronto (using 1914 boundaries) grew from about 13,000 to 73,000, while an increasing number of factories spread over a larger and larger territory.[13] The suburban growth of manufacturing activity was a distinct and important part of urban development in these years, especially in light of competition for factories between the city and the emerging suburbs like Parkdale and West Toronto Junction. Suburban manufacturing activity was fleeting, however, for the city captured this suburban investment in the two major waves of territorial annexation mentioned above. Subsequently, between 1910 and 1914, the growth of manufacturing occurred almost exclusively within the newly expanded city.

The growth of the Toronto urban area, of its economy in general and its manufacturing industries in particular, was stimulated by a number of circumstances. By the 1880s the National Policy's tariff protection for Canadian-made goods had stimulated Toronto's industrial production, resulting in the growth of industries such as agricultural machinery and piano making.[14] Other parts of the National Policy, especially its focus on increased immigration, a national railway network, and banking legislation favoring a few national banks, greatly affected Toronto. The city grew immensely, not only as a manufacturing center, but also as a place of high-order services. In many respects Toronto was almost equal to Montreal, its main competitor, by the time World War I broke out. Both urban areas far exceeded other Canadian cities in manufacturing and had highly diversified urban economies. Even so, the manufacturing sector within these two cities accounted for only about 33 percent of the labor force, which was much lower than the 50 percent or more in some of the second- and third-tier cities.[15]

Manufacturing activity in Toronto increased in volume and changed in terms of organization and technology. In the 1880s the scale of enterprise was still fairly moderate. A few large factories employed workforces of 400 and were largely located in one spot and owned by families or partnerships. Local firms were joined by manufacturers relocating from smaller cities and towns. One of the most prominent firms was the Massey Manufacturing Company, which produced harvesting equipment. By acquiring several firms elsewhere in Canada, the United States, and other countries, Massey became a Toronto-based multilocational organization in the 1890s.[16] Between 1900 and 1914 Canada was gripped by a merger movement that created many large-scale joint-stock companies. Thus corporate capitalism began to replace industrial/entrepreneurial capitalism. The arrival of U.S. branch plants was another expression of the increasing spatial reach of production. The imposition of high customs duties on the import of manufactured goods accelerated the establishment of U.S. firms. By 1913 there were ninety U.S. branch plants located in Toronto—among them Kodak, Coca Cola, Ford, Wrigley Chewing Gums, Canadian Fairbanks-Morse, and Standard Sanitary of Pittsburgh—17 percent of all the U.S. branch plants in Canada.[17] In spite of these changes, however, the size structure of Toronto's manufacturing firms remained extremely diverse. New corporate giants coexisted with workshop-like production units. As we shall see, the large U.S. branch plants (and some of the British ones) were often leaders in the peripheral extensions of the spatial pattern of manufacturing.

Organizational change involved an increase not only in mass production but also in mechanical power as well as changes in factory design. Steam engines were slowly making inroads from the 1860s to the late 1880s. In the 1890s the use of electricity increased, and by the 1910s there was a major boom in the application of electric power. This did not immediately translate into the dispersal of manufacturing. Electricity generated by power companies was first available in the core of the urban area. Factories farther out seem to have run their equipment with electricity generated by their own steam-driven generators.[18] The use of electricity was not automatically associated with single-story factories, which began to appear in the 1880s.[19] These early single-story factories made machinery, and many of them (and all foundries) from that time onward used largely single-story configurations regardless of location. By contrast, the vast majority of manufacturers used multistory buildings in city and suburbs at least until the 1920s.

After the 1880s Toronto manufacturers relied to various degrees on railways, which connected the city to the resources and markets of the hinterland and to the ports of Montreal and New York. Lines fanned out from the center of the city near the waterfront, and a tangential line was placed across the northern edge of the built-up area in the 1880s (Figure 8.2). Factories were established on many of the sites next to railway tracks, and these sites, understandably, were treasured. Suburban municipalities were asked to throw railway sidings into bonus packages and planners discussed bylaws to prevent builders from spoiling valuable sites by erecting houses.[20] Not all manufacturers desired locations adjacent to railways, however. None of the clothing manufacturers and almost none of the printing firms occupied these sites but seem instead to have used central railway transshipment facilities for out-of-town freight. Other industries located far from railway tracks in the 1880s included sash and door factories, planing mills, and piano factories. Freight traffic within the urban area was largely by horse-drawn wagons before World War I. Workers commuted to work on foot or by streetcar. The streetcar network, electrified in 1894, was operated by a private franchise until 1921 and, because the streetcar company refused to serve the expanding urban fringe, public transit followed rather than led urban development.[21] Interurban electric street railways were built in the early 1890s, but they were few in number and failed to link the wide interstitial spaces at the urban fringe. The standard railway facilities were repeatedly pressed into service by manufacturers who ventured too far outward, in order to get workers to the factories when housing was not yet available nearby.

The territory within the city boundaries, which did not change in size between roughly 1800 and 1885, was almost fully developed by the late 1870s, while substantial suburban development took place around 1880. From about 1883 onward, factories appeared in some of these suburbs. This first wave of suburban factories were built at a time of increasingly stiff competition for industrial firms by different municipalities.[22] Thanks to civic boosterism, real estate taxes were waived and a range of bonuses were offered to industrial newcomers. Strained labor relations in Toronto in the 1880s and 1890s also affected the location or relocation of factories. But the suburbanization of manufacturing was slow and sporadic.

A minor exception was the suburban village of Parkdale, located at the western edge of Toronto (Figure 8.1). Parkdale, reputedly a wealthy suburb, offered tax concessions that attracted three factories.[23] In 1889 Parkdale was annexed by the City of Toronto. The only other exception was the municipality of West Toronto Junction, an almost perfect example of a late nineteenth-century industrial suburb.[24] It grew fairly quickly from "incorporated village" (1887) to "town," and then to a small "city" with a population of 12,000. At its founding, West Toronto Junction lay twelve kilometers from Toronto's center and about two kilometers from the city's developed fringe (Figure 8.1). West Toronto Junction had its origin in attempts by the Canadian Pacific Railway (CPR), a latecomer to Toronto, to establish a base in an area crowded by the competition. The CPR acquired railway tracks and land outside the city at an already existing railway junction in the early 1880s, and in 1884 it opened its first railway yard at this locality. A Toronto lawyer, knowledgeable about the railway's interest in the area, purchased land in the early 1880s, hoping to subdivide it and cash in on the demand for housing generated by the railway employees. Although a residential population of 500 was established by 1885 in the unincorporated area, further demand for residential lots stalled. In order to stimulate demand, the landowners tried to attract factories and the CPR's city-based repair shops, but the CPR required an ample supply of water, as did future industrial users. West Toronto Junction's principal landowner attempted to persuade rural York Township's council, of which he had become a member, to spend money on waterworks. When the township refused, he organized the incorporation of the village of West Toronto Junction in 1887. The new municipality quickly borrowed money and installed the waterworks and other infrastructure.

Industrialists benefited greatly from municipal incorporation. They took advantage of ten-year tax freedoms and water supply "at cost." But with benefits came obligations. The industrialists were required to employ an agreed-upon number of workers for at least ten months a year and had to settle at least half of their workers within the municipality. The municipal council also persuaded the CPR to relocate its repair shops, with some 275 workers, from Toronto. The agreement, made in 1890, included an incentive package even richer than that for manufacturers (including free water and a bridge over railway tracks). The city of Toronto fought back and CPR's city repair shops were closed down only in 1907, two years before the annexation of West Toronto Junction. In 1903 the Union Stock

Yards Company moved from Toronto to West Toronto Junction after another incentive package was negotiated. Yet another enhancement of capital accumulation was accomplished with the establishment of West Toronto Junction as a customs port in 1898. The municipality was interested in attracting a branch plant of the Lozier bicycle company of Toledo, Ohio. Lozier wanted a local customs clearance facility in order to speed up the importation of bicycle components needed for a West Toronto Junction assembly factory, and the suburban municipality complied. West Toronto Junction's industrial development policy was quite successful; employment in manufacturing rose from nothing at the time of incorporation in 1887 to 1,842 in 1905.[25]

The strong position some manufacturers enjoyed in shaping the political economy of West Toronto Junction did not produce a "company town," however. By the time it was annexed by the city of Toronto in 1909, West Toronto, as it was called then, was an area of considerable diversity in both economic activity and residential population. After an initial wave of factory arrivals from Toronto and several towns in southern Ontario, West Toronto Junction also managed to attract some branch plants from the United States. The Lozier bicycle company was one of these, and was bought out by a new "combine," organized by Toronto capitalists. This new corporation, the Canada Cycle and Motor Works Company (CCM), made West Toronto Junction the center of a multilocational bicycle company and, beginning in 1905, also branched out into automobile production.[26] Meatpacking companies also began to relocate to West Toronto, among them several older Toronto firms and a Philadelphia company that was later acquired by the U.S.-based Swift Corporation.[27]

A detailed account of the manufacturing establishments in West Toronto Junction in 1908 reveals a considerable range of manufacturing firms in a suburb of about 10,000 residents.[28] Seventeen of the sixty-four firms were corporations, according to the "limited company" suffix in their names. The biggest group of these was in the metals and machinery category (ploughs, industrial transmissions, radiators, bicycles, and automobiles), with others in the wood (pianos, doors and window sashes), food (flour, meatpacking), nonmetallic mineral (bricks and artificial stones), and chemical (paints and soap) categories. There were another seventeen manufacturing establishments with either "company" or "manufacturing" in the firms' names. Brick makers and clothing manufacturers were the most numerous among these kinds of manufacturing companies, but there were also several metals firms. Thirty establishments were engaged in "petty-commodity production." These were small craft-type units such as tailors,

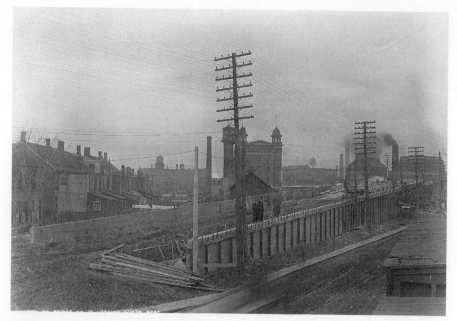

FIGURE 8.3. Multistory Factories and the Weston Road Bridge over the Railway Tracks, West Toronto Junction, 1911.
Source: City of Toronto Archives, Fonds 1231, item 1879. Photograph by Arthur Goss.

milliners, bakers, confectioners, printers and newspaper publishers, blacksmiths, and monument makers. A distinct spatial pattern of production had emerged: the petty-commodity producers were almost all located along the two main streets; the three substantial clothing companies were close to the major residential areas but did not have railway sidings; the piano and other wood products factories were close to the residential areas but each factory had its own railway siding; the metals and machinery factories and workshops and the meatpacking plants had more peripheral locations, with railway sidings for the larger plants but no sidings in the case of the smaller foundries and workshops. The brickyards and paving-stone factories were at the extreme rural north end of the suburban municipality. Even this relatively small suburb thus replicated the locational pattern that can be observed in bigger cities of North America. Almost all of the factories were housed in multistory buildings (see Figure 8.3). The exceptions were the foundries, whether small or large, and the bicycle/automobile factory, which used a combination of multistory and single-story buildings.

A great number of the factory workers seem to have come from outside West Toronto Junction. The 1909 directory lists 278 employees of the Heintzman piano factory, which had 400 employees in 1914.[29] Of these 278 employees, 62 percent lived in West Toronto Junction, while 32 percent came from the city of Toronto and 6 percent from rural areas to the north and west. Harris has shown that commuting into the suburb of West Toronto Junction was common before 1908. In 1896 about 66 percent of the workers employed in West Toronto Junction had come from outside; in 1906 this proportion had fallen to 57 percent.[30] In 1907 large employers like the Canada Cycle and Motor Works Company had complained that workers had great difficulty reaching their factory (which was located at the far end of the suburb). A newspaper reported that 66 percent of the employees in the four largest West Toronto Junction companies commuted outward from the city of Toronto.[31] Clearly it was difficult to match the supply of jobs with the supply of labor and housing, in spite of the policy of the municipal council. In the second half of the 1880s the production of housing outstripped the supply of jobs. But since the industrial policy of West Toronto Junction was so successful, from the mid-1890s onward (until around the time of annexation), the supply of housing could not keep up with the jobs available. Richard Harris has shown that some employers considered supplying company accommodation, but he also speculates that workers may have been hesitant to move to the suburban municipality because employment options were more diverse in Toronto than in the suburb.[32]

Workers may also have been reluctant to move to the suburb because of the level of services provided. Why the city of Toronto annexed West Toronto Junction in 1909, and who benefited most, is not clear. Garland explains the annexation in terms of the threat the suburban municipality posed to Toronto. He claims that, "in essence, the city [of Toronto] could not duplicate the ensemble of infrastructure and services concentrated within the town, nor could it replicate the concessions which capitalists received. Under such circumstances amalgamation was the only viable alternative. It was an amalgamation the terms of which were largely dictated by the suburban municipality."[33]

Although Garland does not specify these conditions, it is clear that the suburban interests gained a great deal. West Toronto Junction had run into financial trouble in the 1890s, when a recession disabled many small land speculators. Since these speculators could not pay real estate taxes, the municipality was forced to sell these tax-indebted lands at fire-sale prices. It also had difficulties with its creditors, since the rapid expansion of infrastructure, partially in the interest of manufacturers and generating little in tax revenue, meant a very heavy debt load. It became increasingly difficult for the suburban municipality to satisfy the expectations of existing and future residents, especially with respect to basic infrastructure such as pavements and sewers. In 1908 the taxpayers voted for amalgamation with the city of Toronto.[34]

The benefits for Toronto are less clear. It could be argued that annexation removed an awkward competitor for new investment in manufacturing. Some Toronto interests clamored for tax incentives and pointed to the example of West Toronto Junction. At the same time, however, manufacturing employment in Toronto grew from 26,200 in 1890 to about 45,000 in 1905. From today's perspective it is hard to believe that West Toronto's 1,000 to 2,000 industrial jobs were a major threat to the city. It seems more likely that there was an imperative for the city to grow. In 1908 a lawyer and former mayor of Toronto explained that the city needed West Toronto Junction because it needed to extend its borrowing powers, which would increase with greater city size. Toronto "had little vacant land on which to expand," he argued, and was amenable to the annexation of the suburb.[35] The annexation brought liabilities for the city but also a diversity of existing manufacturing plants and a huge amount of land for industrial expansion, which took place in the 1910s and 1920s.

The two waves of annexation, 1883–93 and 1905–14, resulted in a large city territory with considerable scope for urban growth (Figure 8.1). The expansion of manufacturing was not only possible on the periphery, inside the new city boundaries, but also in a variety of lands reclaimed by the straightening of the Don River, the draining of marshland, and new landfill in the harbor. In addition, former military reserve lands, which had become only sparsely "colonized" by factories, provided opportunities for the enlargement of manufacturing establishments.[36] The great scope for the growth of manufacturing within one political territory is very instructive, because it shows the possibility of "suburbanization" *within* the "central city." In other words, the large city territory had a distinct central area, an intermediate or inner zone, and most significantly, a periphery or outer zone. This peripheral zone had many of the characteristics of manufacturing in more conventional "suburban" areas. There is no comprehensive, systematic data on the distribution of manufacturing by sectors across these three zones. Only building permit data, which records improvements, expansions, and completely new factories, is of the quality to allow a systematic account.

Table 8.2 provides a snapshot of building permits issued from 1911 to 1914, inclusive, for three broad zones and seventeen industrial sectors.

TABLE 8.2. Value of Building Permits by Industrial Sectors and Three Zones in the City of Toronto, 1911–1914 (thousands of Canadian dollars)

Sector	Central area[a]	Inner zone[a]	Outer zone[a]	Sector total
Printing and publishing	81.0	6.0	13.0	1,237
Loft buildings	65.0	15.7	19.3	724
Nonferrous metals	54.7	0.0	45.3	412
Clothing	53.4	45.0	1.5	681
Electrical appliances	0.0	86.5	13.5	101
Paper products	26.6	72.5	0.8	182
Textiles	23.4	66.5	10.1	137
Leather products	20.0	58.5	21.4	163
Wood products	17.3	57.8	24.9	393
Pianos	10.0	55.4	34.7	108
Miscellaneous	33.6	50.9	15.5	494
Rubber products	0.0	0.0	100.0	198
Transportation equipment	0.0	1.1	98.9	302
Nonmetallic minerals	0.2	9.6	90.2	105
Chemicals	23.0	2.5	74.4	525
Foods and beverages	23.9	30.6	45.5	2,495
Iron and steel products	13.5	40.3	46.2	1,957
All sectors	33.3	29.5	37.2	10,214

Source: Building Permit Files, City of Toronto Archives, selected and aggregated by author.
[a] Zone boundaries, central area: Dundas, Sherbourne, Esplanade, and Front Street west of York, Bathurst; inner zone: outward from central area to Bloor, Don River, Harbour, Dufferin; outer zone: outward from inner zone to city boundaries of 1914.

Proposed investment was almost equally divided between the central area, the inner zone, and the outer zone of the city. Investment in printing and publishing, clothing, nonferrous metals (including jewelry manufacturing), as well as in multi-occupancy loft buildings occurred primarily in the central area. One of the most modern industries of the time, the electrical apparatus industry, and several traditional industries like paper products, textiles, leather products, wood products, and piano manufacturing, were newly established or expanded in the inner zone. Investments in the case of several sectors occurred exclusively or predominantly in the outer zone, the newly annexed parts of the city. There were new or expanded tire plants and other rubber products factories, transportation equipment factories (including car assembly plants), chemical products factories, and brick and other building material manufactories (nonmetallic minerals). Two sectors showed relatively complicated spatial patterns of building permits, since each of the three zones received less than 50 percent of the investment. Both these sectors, foods and beverages and iron and steel products, were heterogeneous, and it is not surprising that different sub-sectors of industries were distributed across the spectrum of zones. However, in these two sectors there were also exceptional industries that showed great locational inertia. Very capital-intensive industries, like brewing and distilling, expanded *in situ*, as did several very large machinery producing factories. These large-scale capital-intensive plants should have moved outward, but for a number of reasons they became stuck in relatively central locations by 1914.

This wave of spatially differentiated investment reinforced earlier geographies of location and reproduced a highly differentiated spatial structure. There was the central business district with a still-thriving garment industry in the old wholesale quarters. There were newly emerging loft districts east and west of the central business district, where printing companies and clothing manufacturers took part in redeveloping older residential areas. There was the inner zone, an area where factories were already loosely clustered in the 1880s and where infill and plant expansions led to increases in production and employment in food, metal products, and wood products industries. Finally, there was the outer zone or new periphery, where the city had annexed areas with factories built in the 1890s and especially in the decade 1900 to 1910. Here, further investment after 1910 led to the predominance of the food and metals industries along with chemicals manufacturers and the emerging rubber products and transportation equipment factories. Beyond Toronto's boundaries there was very little manufacturing activity; the city contained almost all the manufacturing employment of the wider urban area of the mid-1910s.[37]

Manufacturing in the Fragmented Metropolitan Area, 1914–1951

The year 1914 marks an important divide in Toronto's urban geography. After 1914 the city's political boundaries stayed fixed until the formation of the Municipality of Metropolitan Toronto in 1953 changed the political jurisdictions profoundly. After 1914 the suburban share of the urban area's population gradually increased, from about 10 percent to 40 percent in 1951; the suburban share of manufacturing jobs increased from an estimated 3 percent in 1915 to 22 percent in 1951. Suburbanization happened within a framework of strong growth overall: Greater Toronto's population grew from about 500,000 in 1915 to 1.1 million in 1951, and manufacturing employment expanded from about 75,000 to 195,000 (see Figures 8.4 and 8.5).[38]

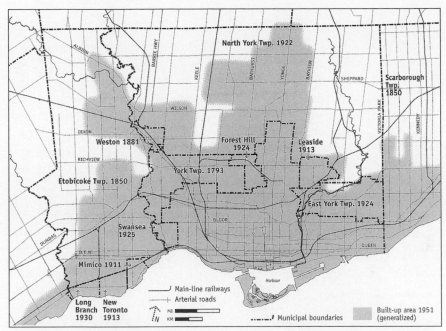

FIGURE 8.4. Municipalities in the Greater Toronto Area, 1951.
Sources: H. C. Goldenberg, *Report of the Royal Commission on Metropolitan Toronto* (Toronto, 1965), 20–21, plate 1; J. Lemon, *Toronto since 1918: An Illustrated History* (Toronto, 1985), 135.

The spreading out of factories over a much larger territory was accompanied by continued growth in Toronto itself. Industrial sectors with a long location history persisted, and new industries engaged in automobile, tire, and aircraft production also started to occupy sites in the central city. Traditional branches of manufacturing, such as whiskey distilling, agricultural equipment manufacturing, and meat processing, expanded by plant enlargements *in situ*. The clothing and printing industries grew in the rapidly infilling loft districts east and especially west of the central business district. The new suburbs beyond the city's boundaries of 1914 were destinations for a number of significant relocatees from Toronto, often firms that were only established between the late 1890s and World War I. A great many factories were entirely new, established by U.S. or British parents or in the context of state-orchestrated wartime production. There were peaks and troughs of production between 1914 and 1951, and the level of suburbanization, although generally increasing over time, fluctuated. Canadian government data make it possible to provide a good quantitative

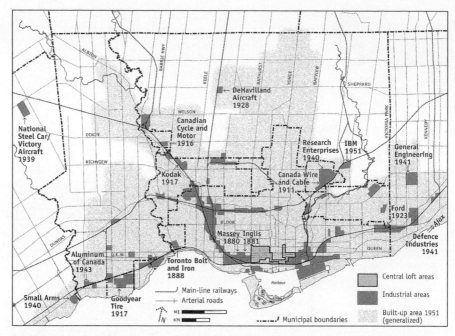

FIGURE 8.5. Industrial Areas and Selected Factories in the Greater Toronto Area, 1951.
Sources: Toronto City Planning Board, Second Annual Report (Toronto, 1943).
Master Plan, including existing industrial areas; Toronto City Directory, 1952; for dates on individual factories, see text and notes.

account of the fluctuating fortunes of central city and suburban manufacturing in terms of employment.[39]

The broad picture outlined in Tables 8.3 and 8.4 is dominated by dramatically changing employment levels in manufacturing. The fluctuations were not quite as strong in Toronto as in the suburbs. At the end of World War I Toronto experienced a historic peak of nearly 107,000 jobs in manufacturing, but by 1921, owing to a very severe recession, employment fell back to prewar levels. The suburbs fared worse; the spurt of expansion during the war was followed by collapse and a receding rate of suburbanization, from 3.8 percent to 2.1 percent. By 1929 manufacturing employment had grown close to the earlier peak in the city and to record levels in the suburbs. The employment share of the suburbs increased to 5.8 percent. Then, in the Depression years, employment was again reduced severely. City and suburbs suffered unequally and the suburban share of employment dipped to 4.3 percent. From the late 1930s to 1943 manufacturing

TABLE 8.3. Manufacturing Employment in the City of Toronto and Incorporated Suburbs, Selected Years, 1915–1951

| | City of Toronto | | Suburbs[a] | | City and suburbs |
	No.	%	No.	%	No.
1915	72,798	99.4	439	0.6	73,237
1920	106,630	96.2	4,238	3.8	110,868
1921	66,708	97.9	1,428	2.1	68,136
1929	102,406	94.2	6,289	5.8	108,694
1933	75,645	95.7	3,420	4.3	79,065
1941	133,099	91.0	13,221	9.0	146,320
1943	156,459	86.1	25,215	13.9	181,674
1946	145,556	90.4	15,532	9.5	160,908
1950	160,063	88.5	20,800	11.5	180,863
1951	151,333	86.3	23,924	13.6	175,257

Sources: For 1915: Canada, *Postal Census of Manufactures of Canada, 1916* (Ottawa, 1917), 188–91. For all other years: Canada, *Manufacturing Industries of Canada* (Ottawa, annual) or *Canada Year Book* (Ottawa, annual).
[a] The suburbs included here are those located in the area delineated in the 1951 Census of Canada as the Toronto Census Metropolitan Area. The individual suburbs were Weston (all years), New Toronto (1920 onward), Leaside (1929 onward), Mimico (1929 and 1943 onward), Swansea (1943 onward), and Long Branch (1946 onward).

employment again increased dramatically in the city of Toronto and even more dramatically in the suburbs, the suburban share reaching a new record of 13.9 percent. In Table 8.4 the more complete suburban record for 1944 shows a substantial buildup of manufacturing in the rural suburbs. Incorporated and rural suburbs together accounted for 21.4 percent of all manufacturing employment in the Greater Toronto area. The shutdown of wartime production facilities again affected the suburbs more than the city, and the slowdown hit the rural suburbs most severely. After 1946, however, the suburban share surged ahead. Between 1950 and 1951 the previously synchronous central city-suburban cycles disengaged; for the first time the suburbs gained manufacturing employment while the city of Toronto lost in absolute terms.

Manufacturing in the suburbs was hampered by the uneven and weakly organized development process outside the city's boundaries. Very few large development companies built suburban towns. Self-building of homes by working-class families was widespread from the 1910s to the 1920s.[40] The provision of urban services, especially paved sidewalks and roads and water supply, was a problem throughout the 1910 to 1950 period.

TABLE 8.4. Central City (City of Toronto) and Suburban Manufacturing Employment, Selected Years, 1944–1951 (thousands of employees)

Year	City of Toronto		Incorporated suburbs[a]		Rural suburbs[b]		All suburbs		Greater Toronto[c]
	No.	%	No.	%	No.	%	No.	%	Total
1944	154	78.6	25	12.7	17	8.7	42	21.4	196
1946	146	88.0	15	9.3	4	2.7	20	11.9	165
1950	160	85.0	21	11.1	6	3.4	27	14.5	187
1951	151	77.5	24	12.2	20	10.2	44	22.4	195

Sources: Canada, Manufacturing Industries of Canada, 1944, 1946, 1950, 1951 (Ottawa, annual).
[a] Included are Leaside, Long Branch (except 1944), Minico, New Toronto, Swansea, and Weston.
[b] Residual area numbers derived by subtracting city of Toronto and incorporated suburbs employment from Greater Toronto total employment.
[c] Greater Toronto is equivalent to the Toronto Census Metropolitan Area of 1951; see also Figure 8.4.

Because of inadequate services, suburbs frequently requested annexation by Toronto, but the city always refused to pay for the process of urban infrastructure expansion. Public transportation was another problem in the suburbs, whereas the city was well served, especially after 1921, when it did not renew the franchise given to a private street-railway company.[41] Interurban railways existed between 1890 and 1930, but there were only a few lines and they did little to facilitate commuting within and between suburbs. In the 1920s Toronto's public transit commission extended some bus lines into the suburbs if suburban municipalities agreed to help finance them. Cars seem to have become widely available in the late 1920s and in the 1930s, but several scholars have argued that they were less used in Toronto than in U.S. cities.[42] Little is known about the industrial policies of the various suburban municipalities. However, bonus enticements by municipalities became more and more difficult and by the 1930s had been practically outlawed by provincial legislation.[43]

The suburbs around Toronto were very diverse. The village, then town, of Weston was broken out of a rural township in 1881; another three municipalities were incorporated between 1911 and 1913, and three more were created between 1924 and 1925. Rural townships not only lost territory to these new municipalities but were also broken down into smaller rural townships. By 1930 the Greater Toronto area consisted of twelve suburban municipalities and the city of Toronto (see Figure 8.4). The suburbs varied enormously in size: in 1921 their populations ranged from 1,000 and 57,000, and in 1951 from 8,000 to 106,000. Harris divided the suburbs into blue-collar (six), socially mixed (three) and affluent (three) municipalities according to occupational characteristics.[44] The twelve municipalities also varied widely according to the number of factories, manufacturing employees, and industrial tax assessment they attracted. Of the six "blue-collar" suburbs, only New Toronto managed to establish a strong industrial base between 1914 and 1951. Of the three mixed suburbs, the fairly well established Weston also had a strong manufacturing component. The "affluent suburbs" showed enormous differences: upper-class Forest Hill did not want factories; middle-class residential areas in Leaside developed next to a large industrial area; and small, middle-class Swansea could boast of a few pre-1914 factories.

The post-1914 suburban zone of Toronto posed considerable challenges to the settlement of people and the establishment of factories. The following sections provide case studies of several incorporated suburbs in order to show the establishment of manufacturing in some detail. Another section deals with manufacturing in the unincorporated rural townships, and a final section discusses the outcome of several decades of manufacturing growth by systematically comparing city and suburbs in terms of industrial sectors.

The town of Weston was an anomaly among the suburbs of Toronto, as it was a rural service town (incorporated in 1881). Its earliest manufacturing establishments were lumber and flour mills and small manufactories. In 1915 Weston had seven manufacturing establishments with a total of 439 employees, including some new, proper factories. The Toronto directories of 1912 and 1915 list the Moffat stove company and its product, the "Moffat Electra" range, under a Weston address. Then, in 1916, the Canada Cycle and Motor Works Company, established in the 1890s in West Toronto Junction, sold its Russell Motor Car subsidiary to Willys-Overland of Toledo (Ohio). Willys-Overland took over motor car production, while CCM, making bicycles and skates, moved to a new single-story factory in Weston, near two railways lines and "adjacent to a pleasant suburban community with excellent schools and moderate real estate values, where employees could own their own houses and could be within easy walking or bicycling distance of their work."[45] Also in 1916, the Massey-Harris agricultural machinery company of Toronto acquired a branch plant in Weston in order to make gasoline engines, and in 1919 the plant started producing tractors.[46] In 1923 Massey-Harris gave up on tractor production and shut the factory down. The facilities were reopened with the advent of World War II, when Massey-Harris received orders for wooden aircraft wings, which became part of Mosquito bombers for the Royal Air Force of Great Britain. In 1941 the factory was greatly enlarged through a major investment by the Canadian government's Department of Munitions and Supply. Shortly after World War II, Massey-Harris closed down production and sold the plant. The Massey-Harris factory in Weston was typical of the larger factories in Toronto's suburbs.

By 1950 Toronto-area maps showed a string of industrial clusters along an axis that stretched from the city of Toronto to the western extremity of the greater metropolitan area. A railway line, an early road along the shores of Lake Ontario, and an interurban street railway, in existence since the early 1890s, had made a string of urban and industrial development possible. Some of this development had happened in a small part of York Township just west of Toronto's Parkdale district, but most of it took place in Etobicoke Township further west. (Parts of Etobicoke Township separated and became incorporated municipalities between 1911 and 1930.)

The first suburban factory along the lake shore axis of Greater Toronto was established in the Swansea section of York Township. In the early 1880s a substantial rolling mill and a bolt factory had been established after a move from the city. They located next to the Toronto-Hamilton railway tracks and seem to have relied on the railway, at least for some time, to transport 500 to 1,000 workers from the city to this suburban location.[47] Farther west along the railway New Toronto had been established in the 1890s. In 1890 a Toronto newspaper reported that a group of Toronto manufacturers had bought 600 acres of land and were in the process of building three factories.[48] The suburbanization process did not seem to have run quite so smoothly before 1914, however. Only three factories lasted (a paper company, a wallpaper factory, and a foundry), while others came and went. Around 1912 a copper and brass rolling mill became a constant fixture. In adjacent Mimico a sewer pipe manufacturer and a fire brick producer became well established. All the factories of reasonable duration had Toronto connections, some of them having moved from the city, others having been established as branch plants. Several had offices in Toronto and almost all the key personnel lived in the city.[49]

After 1914 the industrial landscape of the lake shore suburbs changed considerably. The Toronto-related firms gradually disappeared and large "outsiders" became prominent. In 1917 the U.S. Goodyear Tire Company established a very large multistory factory, which remained Canada's biggest tire factory from 1917 until the 1950s.[50] By the early 1940s the copper and brass rolling mill had been taken over by Anaconda American

Brass, and two other U.S. branch plants—Campbell Soup and Continental Can—had been established next to each other. During World War II the Canadian government helped the Montreal-based Aluminum Company of Canada build a second Toronto plant.[51] This was a hastily put up aluminum foundry that was essential for delivering airplane components. It employed 350 men and 150 women in 1943 and was established in Etobicoke Township, just outside the municipal boundaries of Long Branch. After the war it switched to the production of pistons and other automobile parts. From 1920 to 1951 the number of manufacturing establishments increased gradually from about fifteen to fifty-five in the three incorporated lake shore municipalities. Employment, however, was cyclical. It dropped dramatically—from 3,361 in 1920 to 951 in 1921 in New Toronto. By 1929 the losses had been largely recovered, and employment reached 3,314. The Depression took the number of jobs down again to 2,012 by 1933. Then slow recovery and the demands of the war economy led to a peak of 7,226 in 1944.

The landscape of the lake shore industrial axis had its own peculiar character. Large single-story plants coexisted with substantial multistory ones. Between the large ones small factories and workshops sprang up. Most factories had railway connections that ran along the streets of New Toronto, though very few factories had their own sidings. Factories and residential blocks were only weakly separated in the northern parts of the municipalities. In New Toronto, especially, there was considerable interweaving of factories and very modest, but largely detached, houses. Union halls were visible along several streets. In Harris's typology of Toronto suburbs, New Toronto, Mimico, and Long Branch are classified as blue-collar communities, where a large proportion of the manufacturing jobs were probably filled by local residents.[52] Voting patterns right into the 1990s testify to a high degree of labor solidarity. It is interesting to note that one of the near-revolutionary acts during the Great Depression occurred in Alderwood, a self-built residential area at the western extremity of the lake shore municipalities, when a group of men "detained" the welfare officer and kept him prisoner overnight.[53]

Who was behind the building of these blue-collar suburbs? While Toronto industrialists may have been implicated in establishing New Toronto, and while parts of the lake shore area reflect the "unplanned" creation of suburbs, the presence of the provincial government of Ontario and a railway company must also be taken into account. The provincial government was the first major agent of urban development. It established a large mental asylum at New Toronto in 1890. Around 1900 the provincial government also established a large "reform school" north of New Toronto. These two public-sector institutions attracted a large number of service and professional employees, including many women. In 1906 the Grand Trunk Railway constructed very large railway yards and repair shops, the Mimico Yards, just to the north of the future municipal boundaries of Mimico and New Toronto. While New Toronto comes close to the definition of an industrial suburb or satellite, it also shows some complexity. As in most Toronto suburbs, there was a range of employers, and the residential and social patterns showed a considerable degree of variety.

The manufacturing landscape of Leaside was dramatically different from that of New Toronto. Leaside was a planned town, organized by the Canadian Northern Railway Company.[54] There were no houses between the factories, or even near them, and the railway loading docks were rarely in public streets. Most of the factories were sprawling, modern affairs of one story. Sturdy brick houses predominated in the residential areas, almost all of them with their own driveways and garages. In order to construct maintenance facilities, Canadian Northern acquired land and acted as developer. Through a real estate subsidiary, the company acquired about four square kilometers of land northeast of Toronto in 1911–12. It then commissioned a professional town planner, who submitted plans for Leaside in 1912 and 1913. The plans aimed for an "upper-class" residential area and a separate industrial area of nearly one square kilometer. The industrial area was supposed to accommodate not only Canadian Northern's repair shops and marshaling yards but also many industrial enterprises, which would support the tax base of the "affluent suburb."

The timing was awkward and the planned parallel development of residential and industrial areas did not materialize. For two decades Leaside was an industrial area without a substantial residential population. Instead of a scenario of capitalist exploitation in which local residents were forced to bear the cost of infrastructure, capitalist corporations had to support the town. The first company to arrive, the Canada Wire and Cable Company, shouldered the greatest burden. This company was incorporated in 1911 in the wake of the electricity boom and started production of electric cable in a factory in Toronto. In 1912 it bought land in Leaside and announced the construction of a factory and housing for the workers. When the factory was half completed, however, World War I broke out. The company abandoned the housing project but completed the factory, which immediately began to produce ammunition. Employment soared quickly to some 4,000 workers. Without housing in Leaside, factory workers had to be brought in from the city of Toronto by specially arranged transportation.

Leaside lacked a properly functioning municipal government to provide the infrastructure for manufacturing and housing. The Canadian Northern promoters must have known the difficulties of large-scale urban development, for they sought annexation of their land by Toronto in 1913. The city refused, and Leaside became an independent town with a territory of 4.2 square kilometers and a population of only forty-three. Soon the railway company was in trouble, and in 1919 it became part of the government-controlled Canadian National Railway, which completed the large railway facilities in the southern portion of Leaside's industrial area.

The failure of Leaside's initial developer meant that the Canada Wire and Cable Company, as practically the only major stakeholder, had to incur considerable costs. It helped to finance water mains to be connected to those of Toronto; it had to subsidize and at times completely underwrite transport for its Toronto-based labor force; it built sixty houses to reduce commuting; and it provided the town with electricity generated within the walls of its factory. In the early 1920s Canada Wire and Cable, together with two partners, presented the town with a new fire engine; in case of fire, the factory whistle blew. The first school was housed in an unused company cafeteria building. Between 1924 and 1930 the president of Canada Wire and Cable was also the mayor of Leaside. Labor availability improved greatly in 1927, when a viaduct was built across the huge ravine of the Don River and the flow of labor from the city to the suburb was eased by city-owned Toronto Transit Commission buses. Leaside's reliance on Toronto was manifest not only in the contract with the city's public transit authority but more broadly in the reliance of Canada Wire and Cable on the city's labor force (see also Figure 8.6). In 1931 Leaside's town council praised the town's great advantages as a place for factories and wrote, "Labor conditions at Leaside are ideal. In addition to the Town's permanent 1,200 residents, there is in the section of the City of Toronto adjacent to Leaside a first-class labor market of some 20,000. Largely of native birth with a large percentage owning and living in their own homes, the supply of labor referred to is notably industrious and free from the unrest apparent in many less favoured industrial centers."[55] Here the suburban capitalists were advertising the peaceful labor force of the central city! This may reflect a degree of desperation, but it also shows the relative unimportance of political boundaries, especially those between central city and suburbs, as far as labor conditions were concerned in the first half of the twentieth century.

In spite of problems with access to labor, the Leaside industrial area attracted a steady stream of new factories. The statistical data listed eight

FIGURE 8.6. Large Factories and the Beginnings of Residential Development, Leaside, 1931.
Source: Toronto Public Library, Leaside Branch, photographer unknown.

establishments in 1929, thirty-nine in 1941, and sixty-two in 1951. There was no parallel increase in employment, however. As in other suburban districts of the Greater Toronto area, employment levels fluctuated dramatically. The first boom came with World War I and the first bust at the end of the war, when the 4,000 jobs of the Leaside Munitions Company literally disappeared overnight. In the 1920s employment grew in the cable factory, the adjacent Durant Motor Company automobile assembly plant, and in several paper product and paint factories, reaching 1,816 in 1929. Then, in 1933, the peak Depression year, the auto plant failed and employment dropped to less than half the 1929 level. The last few years of the 1930s saw tremendous employment and residential growth, and by 1942 Leaside had a population of 9,000, though half of the workers in Leaside factories still lived outside the town.

World War II boosted employment to record levels. Leaside was one of the recipients of major government investments in war production facilities. A state-owned company, Research Enterprises Limited, employed 7,500 persons and occupied a twenty-two-hectare site, with a factory floor space of 70,000 square meters, to make radar and other equipment. When the demand for military equipment began to decline, factory jobs were scaled back. In December 1944 Research Enterprises announced the lay-off of 1,000 employees, among them 600 residents of Leaside. The complete shutdown of the company led to yet another phenomenal drop in employment in Leaside, from 13,290 in 1943 to 5,712 in 1946. Recovery in the later 1940s, however, was fairly rapid. The facilities of Research Enterprises were sold to many different companies, including a Toronto radio company (later taken over by Philips Electronics), Honeywell Controls, Corning Glass, and Lincoln Electric Motors. With surviving and new companies making paints, varnishes, paper boxes, and various kinds of machinery in large, generally single-story factories, the modernist suburban character of Leaside was reinforced.

In the rural suburbs, or "townships," of the Toronto area, isolated manufacturers had existed for some time, but large-scale factories were established only in the post-1914 period. Since a good number of the government-sponsored industrial expansion projects of World War II were located in the rural suburbs, these areas merit special attention.[56] These factories are important both in terms of the wartime employment spike in the rural suburbs (42,000 in 1944) and because most of them played a role in post-war expansion and a new wave of the suburbanization of manufacturing.

In two of the rural suburbs, York Township and East York Township, urban development was contiguous with the built-up parts of the Toronto of 1914. Both townships had sizeable populations, 70,000 and 36,000, respectively, in 1931, and both were blue-collar suburbs. Their populations grew in the first three decades of the twentieth century largely through the initiatives of blue-collar workers, who built their own homes. There were few municipal services and few employers. In spite of the ample labor supply, few manufacturers located in these suburbs. The Eastman-Kodak company of Rochester, New York, was one that did, relocating its film-producing plant from Toronto to the Mount Dennis section of York Township in 1917. The Ford Motor Company also relocated its assembly plant from Toronto to East York Township in 1923.[57] Both plants were well located to take advantage of both township labor and the nearby labor force of Toronto. Little is known about whether suburban townships offered bonuses or how employers evaluated labor militancy in these blue-collar suburbs. Considerable militancy emerged in the 1930s, especially in East York Township, where a workers' association was formed and where the newly founded labor party (the Cooperative Commonwealth Federation) received 30 percent of the vote in the 1934 Ontario provincial election.[58]

This suburb became highly accessible to the Leaside industrial area in 1927 through the Leaside Bridge link. Whether the Industrial Commission of the Leaside town council had this labor pool in mind in its 1931 promotional brochure is not known.

Farther away from the built-up areas of the city and the contiguous ring of suburban built-up areas, large factories were scarce but nevertheless significant. One of them was aircraft maker DeHavilland, a British company, which set up a Canadian subsidiary in 1928 and began to assemble wooden aircraft, which came in crates from England, in York Township.[59] In 1929 DeHavilland moved to a new site in North York Township, next to a railway line in open farmland. At the time of the move, DeHavilland had a workforce of only thirty-five, and it is unlikely that the extremely peripheral location was chosen to escape labor unrest. There was no infrastructure apart from the railway line; water supply was a problem, and DeHavilland had to drill its own well. There was no housing and no public transport. DeHavilland's employment grew slowly in the 1930s and then took off, reaching about 300 in 1938 and roughly 2,000 in the early 1940s. The highly routinized assembly of British-designed wooden Mosquito bombers in a large, government-financed plant resulted in an employment peak of 7,000 in 1945. Shortly afterward, waves of layoff notices were handed out, and by the end of 1945 only about 300 workers remained. Recovery was slow, but by 1950–51 the factory was back in business, with new in-house-designed airplanes and a workforce of about 2,600.

Another large-scale factory complex was established in rural Scarborough Township in 1941. The General Engineering Company was a fully owned and operated government enterprise that assembled ammunition.[60] The company expropriated about 80 hectares of farmland and by 1942 had erected 172 buildings, including a school and daycare facility. The munitions plant employed about 5,000 workers, 4,500 of them women. At the end of the war, the complex shut down, and in 1948 Scarborough Township bought the land and converted it to new industrial purposes.

Wartime production also spilled out of the Greater Toronto boundaries of the 1940s. East of Greater Toronto, in Pickering Township, the Montreal-based company Defence Industries Limited built another large ammunitions complex, with government assistance.[61] This complex was literally a company town (Ajax) with its own men's and women's residences, a single-family residential area, a school, a hospital, and many other infrastructure facilities. After World War II the Canadian government converted a substantial part of the land to peacetime industrial uses. West of Greater Toronto, in Toronto Township, the government-owned-and-operated Small Arms Company employed some 1,500 workers; and in the northern part of Toronto Township, at the 1930s Malton airport site, another government factory, Victory Aircraft Limited, assembled the large, British-designed Lancaster bombers.[62] Employment at this aircraft factory reached a peak of 9,600 in 1944 but then dropped to 300 in 1945. A new peacetime owner of the former government facilities gradually resumed aircraft production and employed several thousand workers around 1951.

The World War II boom in manufacturing was very important for the suburbs, especially the outer rural suburbs. Because the factories there were extremely large, these outer suburbs suffered the greatest employment fluctuations of all the subareas of Greater Toronto (see Table 8.4). Employment in the rural suburbs plummeted from 17,000 in 1944 to 4,000 in 1946; in the incorporated suburbs the decline was more moderate, dropping from 15,000 to 12,000, and in Toronto itself the end-of-war decline was very gentle, falling from 154,000 in 1944 to 146,000 in 1946. This was, of course, not the first time that the city, with its more complex industrial structure, held up better than the periphery, with its much less diverse complement of factories. During the latter 1940s, however, and especially around 1950–51, manufacturing employment picked up again in the suburbs— more so in the rural suburbs than in the incorporated suburbs (see Table 8.4). Some of the surviving plants, like DeHavilland and the former Victory Aircraft plant (under new ownership), employed thousands again, largely in buildings erected with wartime government funds. The giant complexes of the General Engineering Company in Scarborough, of Research Enterprises in Leaside, and of Defence Industries in Ajax provided well-serviced land for a new wave of suburban manufacturing. Toronto benefited less than the suburbs, since most of the wartime investment occurred at only one relatively small site (Inglis) close to the city center.

The diversity of manufacturing industries and successive waves of suburbanization have resulted in a differentiated if not complex manufacturing geography. This geography was visible at each stage in Toronto's history, but only post–World War II data allow a fairly systematic and quantitative depiction of the characteristics of central city and suburban manufacturing. Table 8.5 shows the degree of decentralization or suburbanization by sector and also the average establishment size in the suburbs (collectively) and Toronto.

Some of the "modern" twentieth-century industries—automobiles, aircraft, rubber tires, and electric apparatus—had fairly high suburbanization rates (36.4 to 74 percent) in 1951. They also stand out because of very large factories and large differences in plant size between suburban and city

TABLE 8.5. Suburban and Central City Employment Share (%) and Average
Establishment Size (number of employees) by Manufacturing Sector, 1951

	Suburbs		City of Toronto		Greater Toronto
	Employ-ment share[a]	Mean est. size	Employ-ment share	Mean est. size	Total employ-ment
Transportation equipment	74.0%	298	26.0%	78	8,048
Rubber products	46.7	971	53.3	302	6,236
Nonferrous metals	38.0	120	62.0	38	8,876
Paper products	36.9	90	63.1	60	8,542
Nonmetallic minerals	36.8	30	63.2	22	3,392
Electric apparatus	36.4	386	63.6	107	21,234
Chemicals	19.6	29	80.4	38	10,597
Wood products	19.4	16	80.6	16	6,659
Iron and steel	17.1	48	82.9	76	31,055
Textile products	16.7	52	83.3	30	5,643
Printing and publishing	13.6	30	86.4	24	20,098
Leather products	12.7	72	87.3	29	3,419
Food and beverages	11.1	48	88.9	57	25,299
Clothing	2.4	23	97.6	28	22,094
Other	23.8	NA	76.2	NA	13,951
All sectors	22.5	65	77.5	40	195,143

Sources: Greater Toronto/Toronto Census Metropolitan Area: Canada, *General Review of the Manufacturing Industries of Canada, 1951* (Ottawa, 1954), 118–20. City of Toronto: Canada, *The Manufacturing Industries of Canada, 1951*, section 3. Geographical Distribution (Ottawa, 1954), 64–66.
[a] Suburban employment is determined as residual by subtracting city of Toronto employment from Greater Toronto total employment.

locations. Their high degree of suburbanization is not unexpected, nor is the lack of suburbanization of some industries surprising. Printing and publishing, with 86.4 percent of employment in Toronto, and clothing, with 97.6 percent, remained highly centralized. In both sectors establishments were small.

The lack of suburbanization of some of the other sectors, especially food, iron and steel, and chemicals (all less than 20 percent suburban) is harder to explain, especially in light of earlier statements about the spatial dynamics of these industries within the city of Toronto before 1914. Generally, these are capital-intensive industries, but their very nature and their considerable expansion between 1900 and the 1920s may have made it uneconomical for them to move and unlikely for them to add factories in new locations. The food industry included meatpacking plants as major employers. These grew immensely after 1900 and established themselves in West Toronto Junction, annexed by the city in 1909. The rationalization in the baking industry led to the establishment of large bakeries in the outer zone of the city before and after World War I. The breweries and Toronto's major distillery (Gooderham and Worts) were fixed in terms of location around 1880 and immobile until after 1951. The development of the chemicals sector, which included soap and paint factories as major components, mirrored that of the food industry. Large investments were made in Toronto's outer zone before and after World War I, and there may also have been some functional interdependence with the meatpacking industry that made suburbanization unlikely. The iron and steel sector, which included machinery and agricultural implement factories, were also fixed in terms of location. When overall employment numbers are considered, the situation of large and leading-edge companies like Massey-Harris and Inglis may help to explain the lack of suburbanization. (It is also interesting that average establishment size in the metal products industry was considerably higher in the city than in the suburbs.) Agricultural equipment manufacturer Massey-Harris several times considered moving out of its very central location in Toronto, but each time it was able to acquire more land or reorganize its production arrangements *in situ*.[63] The sectors that remained highly centralized were those with the largest employment numbers (Table 8.5) and this circumstance explains the high rate of overall employment centralization.

The preponderance of larger factories in the suburbs suggests that the vertical integration of production might have been an important factor in generating the pattern of spatial differentiation. Little is known, however, about the functioning and the linkages of manufacturing in the Toronto area before 1950. Some glimpses of conditions in the World War II aircraft industry suggest large plant size, capital-intensive production, and extremely peripheral locations were not necessarily predicated on vertical integration. DeHavilland's Mosquito bomber production reputedly involved 200 subcontractors.[64] Thirty of these were located at multiple sites: in Toronto suburbs (except the rural suburbs), in Toronto (where DeHavilland built the fuselages), in other Canadian cities, in the United States, and in Great Britain. We have no information about how the various components got to the assembly plant, except that the wings came by motorized truck (the location of the plant next to the railway tracks must also have been important). There is skimpy evidence about other linkages between factories and the functioning of industrial districts. In the 1880s, and presumably into the 1910s and 1920s, an industrial district existed in the eastern zone of Toronto, where linkages involved a distillery, breweries, slaughtering, meat processing, and leather and soap manufacturing. No similar industrial complex grew up, however, in the meatpacking cluster in the western area of the city after the annexation of West Toronto Junction (though two or three meatpackers located there in 1914 had attracted fertilizer factories). Daniel Hiebert describes the tight interrelationships in the centrally located clothing industry.[65] Less is known about printing and publishing, but we do have anecdotal evidence of the many connections to centrally located news gathering, centrally located consumers of printed products (legal profession, banks, governments) and the importance of knowledge workers such as writers, designers, and artists.

Harris and Bloomfield, and Harris have partially documented the relationship between production sites and worker residences, but the evidence is not easily summarized.[66] In the case of some districts (e.g., New Toronto) there seems to have been reasonable proximity between residence and workplace. Some blue-collar suburbs required long commutes, however. There was also a lot of reverse commuting in the opening years of the century to the outlying plants (in the case of West Toronto Junction), in the 1920s and 1930s (to Leaside), and during World War II and afterward to the large, isolated factories. Housing the workers of the isolated plants was always a problem, and mistrust of isolated factories may help to explain the rather slow suburbanization of industrial labor. Toronto's reluctance to pay for suburban infrastructure, including transportation, and the relatively good services within the city not only retarded suburbanization but created a flexible labor field in the city. It is reasonably safe to say that the Greater Toronto area worked fairly well as an integrated labor market without a strong multinodal, cellular articulation for most of the years between 1914 and 1951.

Summing Up

One strand of the literature on the suburbanization of manufacturing emphasizes the role of industrial capital in driving the pace and character of the process. In this view, the industrialists cooperate with real estate interests and suburban municipalities in order to provide for smooth urban development and capital accumulation. The suburbanization process in Toronto does not entirely fit this picture. There were instances when industrialists, real estate developers, and municipal governments collaborated successfully, as in the case of West Toronto Junction in the 1880s and 1890s, but this was the exception. Collaboration did not take place in other suburbs in the 1880s and 1890s, and it did not take place between 1900 and 1950. In Toronto the suburbanization process, and especially the suburbanization of manufacturing, unfolded unevenly over time and space. The relatively few large manufacturers who ventured ahead of the spread of the developed area usually paid a considerable price in terms of ease of operations. At the same time, Toronto in the early twentieth century offered space for industrial expansion within its boundaries. It also offered a compact and well-functioning public transit system, especially after 1920, and there is no evidence that there were major differences in labor militancy between the city and the suburbs.

The difficulties for manufacturing in the suburban realm differed according to economic cycles and the timing of the two world wars. While downward cycles in the economy swiftly led to downsizing of suburban manufacturing employment, booms increased the number of manufacturing establishments and especially the number of jobs. The two wars were especially important for the expansion of suburban manufacturing. Although governments did not engage in any comprehensive urban planning, government resources poured not only into plants but also into housing, transportation, and daycare.

In Toronto's case, the beginning of the suburbanization of manufacturing cannot be dated precisely. Only the creation—and extinction—of municipal governments can be chronicled exactly. The spread of manufacturing has been a continuous process dating from at least the 1880s (and no doubt from the decades before that). Throughout the history of the Greater Toronto area there have been centrally, not-so-centrally, and peripherally located factories. At times the imposition of political boundaries made a number of these factories "suburban." It has also been shown that the spread and suburbanization of manufacturing was not predicated on the advent of single-story factories, the availability of electric power, or the rise of mass production. Suburbanization also predates the advent of the motor truck and the automobile.

Some authors have reflected carefully on the spatial differentials arising from the spreading out or "decentralization" process. They have argued that large-scale material-intensive production and/or capital-intensive manufacturing is prone to decentralization, while small-scale labor-intensive production typically stays in central locations. In Toronto, some large-scale capital-intensive production units were always part of the advancing urban fringe, even in the 1880s and earlier. Some large-scale capital-intensive production facilities, however, did not move outward to the suburbs. The failure to relocate was, at least in some instances, merely random. Several systemic factors also came into play. The opportunity to acquire land in the central parts of the city made relocation unnecessary for some factories. Investment in highly specific manufacturing buildings, as for instance in the brewing and meatpacking industries, caused capital-intensive production to remain in central or intermediate locations. Small-scale labor-intensive operations were not only located in the central parts of the urban area but also shared the industrial areas of the suburbs with the large factories.

Finally, commuting distances between home and workplace played a role. Literature, both academic and popular, is fond of the notion of "urban villages." It is often assumed that workers lived close to the factory gate. As far as we know, things were rarely so neat in Toronto. At certain times and in certain places (New Toronto, for instance), most workers do seem to have lived close to their factories; but the greater parts of Toronto have seen fairly flexible connections between home and work. Within Toronto itself this flexibility may have been fostered by a high degree of spatial compactness of urban development and the existence of a streetcar network. On the other hand, the journey to work undoubtedly caused great hardship for those living beyond the city's boundary, especially in the rural townships.

There is still a great deal that we do not know about Toronto's industrial geography. For example, the size structure of manufacturing establishments is not fully documented, and it is not clear what function the medium-sized and, especially, the small establishments performed in the suburban areas. Were they part of integrated manufacturing clusters and industrial districts? Much more needs to be done on linkages between firms in Toronto. We need to know more about the means of transportation of goods between factories, about the journey to work for different time periods, about the unionization of labor in the city and suburban plants and the geography of labor militancy. While bonuses generally declined in the first half of the twentieth century, nagging questions remain about several large factories that were established just outside the limits of incorporated municipalities in the rural townships.

One thing is certain: after 1951 the city of Toronto steadily lost manufacturing employment to the suburbs, because the conditions for manufacturing changed dramatically in the suburbs with the formation of the municipality of metropolitan Toronto in 1953.[67] The new metropolitan government swiftly expanded roads, sewers, water mains, and public transit, thus creating ideal conditions for development and for the construction of factories in the suburbs. At the same time, several booms in office development and the growth of universities, medical facilities, and cultural venues led to a restructuring of land use and new demands for residential accommodation in the city. In the second half of the twentieth century the pull of new suburban infrastructure was aided by a push created by competition for space in the inner city.

9

"Nature's Workshop"

Industry and Urban Expansion in Southern California, 1900–1950

GREG HISE

The industrialization of [Los Angeles] was not unwelcome; in fact, it was to a considerable extent deliberately planned and cultivated.

—John Parke Young, quoted in G. Robbins and L. Tilton, *Los Angeles: Preface to a Master Plan* (Los Angeles, 1941), 61.

n 1949 a journalist visiting Los Angeles on assignment for *Fortune* magazine drove a four-mile, west-to-east transect along 190th Street. Beginning in Redondo Beach, the writer passed through Torrance, the city of Los Angeles, Carson, and out into unincorporated county land, observing a "truck farm landscape with acres of new factories." Factories sprouting in farmland, a common trope, played off readers' conceptions of Los Angeles as both a place of agricultural abundance and the great boom city of the twentieth century. These widely shared notions of southern California provided the counterpoint for this reporter's understated assessment of the city: "The most remarkable thing is that in the Los Angeles of 1949, [190th street] is utterly unre-

Reprinted with minor revisions from Greg Hise, "'Nature's Workshop': Industry and Urban Expansion in Southern California, 1900–1950," *Journal of Historical Geography* 27 (2001): 74–92, by permission of the publisher Academic Press.

markable. . . . In and around Los Angeles, in the space of a very few years, there has grown up one of the great industrial complexes in the world." Continuing north into Vernon, this visitor found "absurd Moorish factories of the 1920s around the corner from narrow, dark streets where the high walls and catwalks seem to have been transplanted from New England," a juxtaposition that seemed even more fascinating given its proximity to "some of the most exciting industrial architecture in the United States in the Santa Fe railroad's carefully zoned Central Manufacturing District."[1]

Certainly this reporter's account of industry and urban expansion was intended to capture firsthand the magnitude and nature of change in southern California during World War II.[2] From this vantage point the topic is exceptional, or at the very least novel; it cuts against the grain of popular depictions of postwar Los Angeles as the "biggest darn collection of suburbs in the world." If we read it another way, in this case thematically, the *Fortune* essay underscores the pervasiveness of accepted suburban narratives. Stock accounts of suburbanization, especially those devoted to the post–World War II era, feature a number of protagonists and a variety of sites. The majority employ four narrative conventions: the settings depicted represent a new landscape order; once outside the "city," specific location is secondary if not immaterial; the vast array of individual development projects that constitute a particular suburb are staged and completed without the benefit of coordination or deliberate planning; and the resultant physical and social patterns display an increased degree of homogeneity relative to past trends.[3]

Consider the *Fortune* excerpt with these conventions in mind. The overall tone of the piece, what is intended to interest readers, is the newness of industrial Los Angeles. The current crop of factories, completed "in the space of a very few years," are presented as seemingly random, with no discernible logic that might explain why manufacturing facilities were located in particular locations; in other words, the expansion does not appear to have any limits or boundary. What is discovered, in essence, is industrial sprawl; over time and across space this sameness or repetition becomes "utterly unremarkable." In the final analysis, industries are undifferentiated, perhaps interchangeable; plants might be found in any number of similar locations throughout the region.[4]

The novelty, then, in the *Fortune* report is simply the object of investigation. The structure is generic, the explanation formulaic, and the author rehearses a well-known plot. But is this the only plot? Is it the most appropriate? I think not. The planned dispersion of industry and associated land uses has a deep history that can be traced to the mid-nineteenth century.

Mixed-use districts had been the norm in North American cities, rather than a new landscape order as mid-twentieth-century analysts imagined.[5] Historical analysis of industrial geography in Los Angeles contributes to this reinterpretation because it requires a reworking of each of the conventions associated with the standard suburban narrative. The turn of the century serves as a benchmark for this examination of the dynamic shifts in industrial sectors, production techniques, and the creation of new districts throughout the region during the decades leading up to the World War II defense emergency. The 1920s, a generative decade when civic elites, entrepreneurs, and workers fixed the coordinates for industrial development in southern California, will be the focus of this essay.

In Los Angeles, as elsewhere, industrialists, working in concert with land developers, realtors, design professionals, and other city builders, helped shape the precise pattern of urban expansion, contrary to the standard suburban narratives that mythologized urban expansion as an unplanned, potentially limitless sprawl. The Los Angeles Area Chamber of Commerce (LAACC or the Chamber) and other boosters touted Los Angeles as "Nature's Workshop, [the] city where nature helps industry most."[6] But this conceit, that manufacturers would enjoy the gift of a benign climate and sunshine, obscures the role conscious planning, both comprehensive and incremental, played in creating the conditions necessary for eventual industrial expansion. In Los Angeles, industrial, commercial, civic, and residential development was concomitant and complementary. Previous interpretations of this expansion treat industries as homogeneous, yet they are distinguished by type and by the particular requirements of firms within a specific sector. These variations have implications for locational choices and prior or subsequent transformations in the urban landscape. The critical variables include land availability and acquisition, access to transportation, water, and power, proximity to ancillary firms and services, and a stable and adequate workforce.

This investigation of industrial Los Angeles points to a larger, more synthetic claim; it challenges prior accounts of industrial development in the United States and Canada that chronicle a series of successive industrial regimes, each with its associated spatial and social orders, and presume the inexorable march of progress in which each emergent set of events eclipses preceding conditions in a continuous advance. Put simply, in Los Angeles, the nature, timing, and pace of industrial development combined to collapse these putatively distinct epochs to such a degree that the standard theories pertain only if they are elastic enough to explain states that, in Los Angeles, were not successive but concomitant.

Industry, Planning, and the Comprehensive Dispersion of Urban Functions

Under the auspices of its industrial department (initiated in 1918), the LAACC actively promoted the reputed advantages of southern California for manufacturing and aggressively courted investors and financiers who might capitalize industrial startups or the expansion of existing firms and plant.[7] Touting a benign climate, abundant, low-cost hydroelectric power and petroleum-derived fuels, labor costs below the national average, a bungalow-dwelling workforce, and the open shop, boosters lured investment capital to Los Angeles.[8] One of their strategies was to entice East Coast and midwestern corporations such as Ford Motor Company, General Motors, and Willys-Overland to select Los Angeles as the site for satellite production facilities or branch plants. Promoters anticipated a secondary benefit. Branch plants, they reasoned, would create the impetus for investment in resource extraction, basic industries, and ancillary production.

This branch plant expansion, as well as the growth of locally owned firms, did not produce a generic industrial Los Angeles.[9] Firms set up shop, hired workers, and manufactured products for a local market and later for export. The timing of this development and the specific urban patterns that resulted raise a set of interlocking, or overlapping, concerns such as location, taxes and municipal ordinances, labor requirements, and intra-metropolitan competition. The growth process depended on a temporally contingent intersection of variables that included, but were not limited to, the value of land, the availability and cost of financing, infrastructure, zoning, and the individual choices made by workers and their families. To cite just one example, developers and industrial realtors capitalized on variations in tax structures between municipal jurisdictions as fundamental differences that made property in one location more attractive than the parcels available in another. At the City Industrial Tract, which straddles a city-county boundary in East Los Angeles, the Walter Leimert Co. advertised the tax advantage industrialists would gain by locating in the county. In this case, position translated into a savings of $1.26 per $100 dollars of valuation. Just as critically, according to the company's marketing material, "Industry located in the county is freer from governmental restrictions." Leimert worked the municipal tax and infrastructure incentive angle continuously and creatively. He located the housing for this mixed-use, workplace-and-residence development on the city side of the jurisdictional divide and trumpeted the advantages this location afforded lot buyers in advertisements for City Terrace.[10] At the same time, Leimert and his contemporaries realized that the parcels designated for industry on the county's land-use maps were not equivalent. William French made this point as part of a 1926 study of locational factors for manufacturing plants. French inventoried available land and property values in seventeen incorporated communities within Los Angeles County and found "unimproved sites being offered as low as $400 per acre and improved land in outlying districts being sold for a high as $35,000. Some of the land in Los Angeles is valued as high as $60,000 to $100,000 per acre. The range of prices does not represent the range in industrial values of the sites offered, because some of the cheaper sites are better situated than others held at higher prices."[11] Some analysts argued that too much land was zoned for industry. Following a survey of "The Industrial Land Situation in Los Angeles" completed in August 1925, economic consultants and civil engineers Eberle & Riggleman claimed this was a grave danger. They urged manufacturers to pay close attention to distinctions in infrastructure improvements (and to the bearing capacity of soils) and voiced a general concern regarding "indiscriminate dispersion" that would militate a desirable "cohesion and interdependence of industrial facilities." These consultants also pressed for regional cooperation, arguing against the internecine competition that characterized industrial promotion during the 1920s. They suggested that municipalities specialize in a "particular type of industry in which they are most interested and to which their geographical situation may be best adapted." Their examples included furniture manufacturing in Inglewood and heavy metal products and machinery manufacturing in Torrance. Under such a plan, competition would be eliminated and a more effective operating area established for each industry.[12] It is equally clear that the principals involved in creating industrial Los Angeles recognized that all manufacturing was not equivalent. In its prospectus for City Industrial Tract, the Leimert Co. highlighted a set of restrictive clauses written into the property deeds. These regulations were not intended to protect residents living in City Terrace. Rather, they were designed to protect the purchasers of industrial property against the intrusion of "obnoxious or detrimental types of industry," including chemicals, fertilizers and insecticides, stockyards and fat rendering, and excavation for brick, tile, or terra-cotta.[13]

Industrial developers also noted the distinctions that characterized industrial sectors, especially those factors that influenced locational decisions and the types of manufacturing facilities that were constructed or leased, as well as the differential timing of a sector's emergence in the region. Each of these factors informed the geography of manufacturing in

Los Angeles. A comprehensive assessment would consider the ways that industrial development intersected with the plotting and construction of residential districts, commercial districts, and the location of services, cultural institutions, and civic centers—in other words, the overall pattern of city-building. In Los Angeles, these different types of development were concomitant. This was not an accident. City builders of all types understood that these seemingly discrete types of development were contingent, and their coincident enterprises resulted in an expansive recasting of the city during the first half of the twentieth century. The critical finding is that there were three interrelated but qualitatively distinct types of industrial zones in Los Angeles during the period under consideration (1910–35); two of the three were emergent, but each was vibrant and expanding (see Figure 9.1).[14]

The East Side Industrial District

As recently as ten years ago, in discussing the future of Los Angeles, the most enthusiastic boomer would never claim that it was ever likely to become an important manufacturing center. . . . Today, in the main city thoroughfares, he beholds the business bustle of New York, along the old river bed a small Pittsburgh of factories and workshops, away toward the west and south, a sweep of parklike lands covered with the homes that have made Los Angeles world famous.

—Los Angeles Times, 1 January 1905[15]

This is a surprising characterization of turn-of-the-century Los Angeles, a city that has long been heralded as a self-conscious antipode to East Coast and midwestern industrial cities. In fact, the reporter's vainglorious statement underscores the marginal stature of Los Angeles relative to New York and Pittsburgh. Nevertheless, the comparisons served as a call to action; Los Angeles elites knew their city lacked the bustle of commerce and the manufacturing it needed if it were to claim a rank higher in the urban hierarchy and become more than "a small Pittsburgh." Angelenos knew the area of factories and workshops along the river by another name, the East Side Industrial District, a linear, mixed-use zone that paralleled the trunk and spur lines of the national railroads. In many ways the name was a misnomer. Although there were a number of manufacturing concerns, the area was a polyglot landscape similar to the central areas in cities like New York and Pittsburgh. In Los Angeles, this district housed an extraordinary diversity of land uses, activities, and people; foundries, boiler works, patternmakers' shops, iron works, stores, restaurants, saloons, and residences that ranged from single-family dwellings to apartments and furnished rooms.[16]

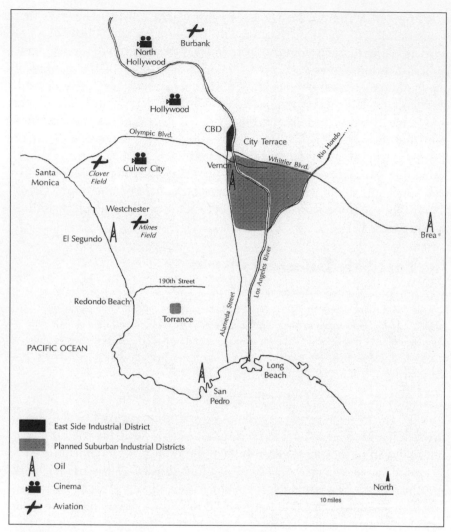

FIGURE 9.1. The Industrial Districts of Los Angeles.

Most of the manufacturers in the East Side Industrial District produced for a local market, and Angelenos owned most of the firms. The Los Angeles Soap Company, representative of a medium-sized to large firm with deep roots in the district, originated in the 1860s when John A. Forthmann moved his soap business to a greenfield site adjacent to the San Pedro, Los Angeles and Salt Lake railway. Over time, the product line expanded to seventy-five brands, output increased, and Forthmann incorporated. In

1898 he secured a parcel on East First Street and contracted for the construction of a brick glycerine recovery plant with business offices on the second floor. The company repeatedly acquired neighboring parcels and conducted an ongoing building campaign that culminated in plant and yards occupying twenty acres, a workforce of 500 employees, and bar, laundry, and toilet soap shipped "east to the Mississippi River with trade out of Los Angeles harbor to Hawaii, the Orient, and the Gulf of Mexico."[17] The district also has been a center for apparel, furniture, and food processing, as well as a prime location for jobbers, wholesalers, and associated warehouses. One mile south of Los Angeles Soap, on a twenty-seven-acre parcel, the Union Terminal Company developed a two-million-square-foot produce center. This facility, sited at the "western edge of the industrial section and the eastern edge of the retail section," provided switching and private rail tracks and immediate access to Pacific Electric freight cars, with a connection to the Southern Pacific depot at Stephenson Street between Third and Fourth streets. The first phase of construction, begun in 1917, included three six-story buildings with two million square feet of leasable warehouse space, one million cubic feet of cold storage, and a set of two-story market structures forming an open-air yard for wholesale trade. The development company patterned this project, the largest of its kind west of Chicago, after the Bush Terminal Company's facility in New York and a similar project completed by the Piedmont Lines in Charlotte, North Carolina.[18]

Los Angeles Soap and the Wholesale Terminal were representative of the larger firms and facilities in the East Side Industrial District. The soap company developed its site and plant in an accretive fashion; the Union Terminal Company controlled a significant parcel and planned for and then constructed an efficient ensemble of up-to-date warehouses and offices. The majority of firms in this district occupied small lots and produced irrigation machinery, chemicals and pharmaceuticals, machine parts and ornamental ironwork, paint, furniture, and bakery, confections, and other household goods. Reformer Bromley Oxnam noted in 1925 that the area around the Macy Street School "is gradually being changed into industrial sites. The land in this section is held by more than 120 owners, most holding one to six lots."[19]

When William French surveyed property holdings and land value for his 1926 study he found that, with the exception of two or three parcels, industrial property in this part of the city was either built up or valued at a price prohibitive for manufacturing and concluded that "this tendency has forced some industries to move from the district in order to expand."[20]

Although French and other observers noted the perceived shortcomings that might restrict industrial development in the East Side Industrial District, this did not reflect a consensus; investors and firms continued to locate there. In 1919 the California Commission of Immigration and Housing (CCIH) reported that just over 33 percent of the manufacturing concerns in the city were located in this district, and the east side remained a desirable location and a robust and expanding mixed-use zone throughout the 1920s and into the postwar years. A classified advertisement from 1926 in the *Los Angeles Times* offered a parcel on Alameda Street for long- or short-term lease. The lessee presented the land as "industrial acreage ... suitable for graders' or contractors' headquarters, livestock market, dairy, poultry, factory, lumber yard, race track or amusement arena." During the defense emergency, when the demand for floor space exceeded supply, aircraft firms that had dispersed out from this downtown adjacent zone and recentralized in specialized districts around outlying airfields returned some of their operations to loft buildings in the East Side Industrial District. More recently this part of the city has emerged as an "incubator site for immigrant entrepreneurs."[21]

Industrialists took many factors into account when deciding whether to remain in the East Side Industrial District or move out to City Industrial Park, the Union Pacific's Metropolitan Warehouse district, Vernon and the CMD, Torrance, Culver City, or other outlying locations, but the critical factors were zoning and workers' efforts to improve shop floor conditions and secure increased authority in their negotiations with management. Los Angeles residents approved zoning in response to the noise, filth, and general hazards associated with brick making, slaughtering, fuel generation, and other noxious industries.[22] The city council drafted a series of legal regulations that spelled out where manufacturing could locate within city boundaries, beginning with an ordinance restricting certain industrial uses in a residential area in 1904. In 1908 and 1909 municipal authorities approved statutes that parsed the city into two residential and seven industrial districts. The next year an ordinance designated as residential all city land not falling within the industrial districts. Other incorporated communities followed suit and assigned land for industrial development. Although designed to protect single-family housing, this legislation promoted dispersed industrial clusters that encouraged manufacturers and developers of these tracts to plan for working-class housing and services in close proximity to employment.[23]

In effect, this type of restrictive legislation served industrialists' interests because it set aside entire swaths of the city for manufacturing. In 1928 the Chamber's Manufacturing and Industries Committee reported to the board regarding an "ideal plan worked out by the Trackage Committee, under the direction of County Regional Planning Engineer W. J. Fox, to coordinate the lands of various owners in a given locality for the purpose of industrial development and obtaining trackage connections, especially where the tendency was to subdivide such property into residential lots." The plan and the Chamber's enthusiastic support for it underscore the ways industrialists capitalized on planning initiatives. In this case they helped craft the planning agenda through cooperation and coordination with one of the county's senior planning officials. The Chamber's position on the zoning issue was expressed tersely during a 1922 board meeting. When a director asked if the Chamber had gone on record "in favor or against the policy of zoning," a colleague replied: "It is a child of the Chamber. We started it."[24]

Relations between labor and management had complex effects on industrial location decisions. Metalworkers, one of the more organized trades in Los Angeles, had the most impact. During the first decade of the century these workers walked off the job en masse on three occasions (1902, 1903, 1910) in order to secure a nine-hour workday, overtime pay, and fixed hourly wages. Firms engaged in the production of pipe (for irrigation and oil), in sizing and preparing structural steel (for bridges and buildings), and in the manufacture of pumps, lifts, and boilers had concentrated in the East Side Industrial District, a majority north of the Plaza. Fred L. Baker and John Llewellyn (proprietors of Baker Iron Works and Llewellyn Iron Works) organized management in a Founders' and Employers' Association (FEA). This group's board of directors included representatives from the region's twenty-five largest concerns; collectively these employers led a militantly antiunion crusade. Baker and Llewellyn repeatedly hired nonunion workers to replace strikers. Following the 1903 walkout, Baker demanded that workers sign an affidavit of non-membership in any union. Both men lent their support to the 1910 employers' initiative to criminalize picketing, a response to a strike that drew 1,500 metalworkers off the job and threatened production at all FEA-member plants. Llewellyn and Baker Iron Works, as well as two of the firms' project sites, were targets in the series of explosions that racked Los Angeles that year. The primary evidence does not provide definitive links that connect a decade of labor activism and the dynamiting of the *Los Angeles Times* and Llewellyn Iron Works causally with the decision of firms like Llewellyn to establish satellite facilities outside the East Side Industrial District. Still, Llewellyn's decision to locate in Jared Sidney

Torrance's eponymous "model industrial city" cannot be explained solely by reference to the purported advantages of a greenfield site with "protective zoning for industrial use," "non-speculative ownership in large tracts," and low land costs.[25]

Planned Suburban Industrial Districts: Torrance, the Eastside, Vernon

It will be our own fault if we do not direct our city into proper channels of growth. This means that industries must be scattered throughout the whole metropolitan district. It must be decentralized. Such decentralization will make for better living conditions and better citizenship, as well as for cheaper overhead costs.

—A. G. Arnoll (1924)[26]

Torrance is representative of a second type of industrial landscape: dispersed greenfield sites engineered with up-to-date infrastructure and services, intended for firms whose production relied on the beltline and other technologies for the mass production of metal fittings, refinery equipment, tires and tubes, and automobiles. These goods were crafted and assembled for a regional market and increasingly for export throughout the West and across the Pacific. Yet Torrance had exceptional qualities: the control of a single corporation, a master plan and guidelines for a comprehensive satellite community, and a high degree of coordination by the developers. In 1911 Jared Torrance, an entrepreneur who came to Los Angeles from Gowanda, New York, and made his fortune in the southern California trinity—railroads, real estate, and oil—announced plans for an industrial city. The timing, relative to the bombings and the mayoral bid of socialist lawyer Job Harriman, was not accidental. Torrance incorporated the Dominguez Land Corporation with financier Joseph J. Sartori (Security Pacific Bank), purchased 2,800 acres in southwest Los Angeles, and hired F. L. Olmsted Jr. and Irving Gill as designers. Olmsted's site plan set out a transit gateway, civic center with theater, public library, and linear park, all leading to quarters for detached workingmen's cottages. A declaration of "reservations, restrictions, conditions, covenants, charges and agreements" issued by the Dominguez Land Company delineated land uses precisely by type; a set of "race restrictions" confined "any person other than of the white or Caucasian Race" to the "foreign quarters." Industrial development was piecemeal until 1916, when Dominguez Land donated a 125-acre parcel to the Pacific Electric Railway for its construction and repair yard, a predictable gambit in the internecine politics of urban growth.[27]

City boosters, developers, and planners understood industry and city building in a manner similar to that of Jared Torrance: they believed industry should be zoned for discrete segments of the city and segregated by type; they endorsed and encouraged development that surrounded production landscapes with a complement of residences, services, and community institutions; and most developers and industrialists agreed with Torrance regarding fixed boundaries segregating the "races." Nevertheless, all industry did not disperse and recentralize in these new districts, all outlying districts planned and zoned industrial did not become sites for production, and no consensus existed regarding the optimal location for local firms or companies either relocating to the region or setting up branch facilities. In fact, a number of industrial districts and their residential counterparts sited outside the city's municipal boundaries were planned, serviced, and marketed during the 1920s but remained sparsely developed up to the defense emergency.

During a January 1922 meeting of the Chamber of Commerce, these property holders and boosters engaged in a rancorous debate regarding the relative merits of "opening up" a section of San Pedro Street adjacent to the central business district (CBD) for industry, a move, they duly noted, guaranteed to antagonize voters on Boyle Heights. On one level, this was a debate concerning where Westinghouse should build its first Los Angeles facility. The proposal before the board called for a site at the corner of Ninth and San Pedro streets. On another level, this resolution was a proposal to transform San Pedro into a "wholesale district." As Chamber president Weaver noted, "all those who do not believe in it as a wholesale section and who do not want [the construction of additional] grade crossings will be in opposition." Those opposed argued for a Vernon location, where there was "plenty of vacant ground served by three railroads only twenty minutes from 7th [Street] and Broadway." At this inflammatory session, the opposition made unsavory comparisons to New York, Pittsburgh, and St. Louis, while Weaver argued that he was all for "giving Los Angeles the advantages those cities have." Though united in principle, these antagonists drew precise distinctions between different types of industry, the exact needs of particular firms, the appropriate location for various manufacturing activities, and, most critically, the optimal pattern of land uses in Los Angeles. The intra-metropolitan geography of industry, as the majority defined it, had warehouses and jobbers serving downtown commercial and retail with production segregated to outlying districts such as Vernon, the Union Pacific's Metropolitan Warehouse and Industrial District, and the Southern Pacific's tracts in present-day Commerce. This reflected patterns of land ownership by board members but also concerns about property values in the CBD and the depth of anxiety during the 1920s regarding transportation and congestion in that part of the city.[28]

Residents also perceived the city dichotomously. In 1924 a contributor to the *Times* stated boldly that in order to understand Los Angeles it was necessary to motor south through the manufacturing district. "The rest of the city, from the winsome foothills to the glittering beaches, when viewed alone does not convey an adequate idea of the true situation. Along with all that it has been in the past, this city is now an industrial entity." A drive from the east side southward would bring into view a "stage for the newest and biggest act in the great Los Angeles drama." Here the air "is filled with industrial haze and queer smells, huge trucks trundle along paved thoroughfares. [This] is the new city; it is not amusements or tourists, it is industrial production."[29]

The Chamber expended considerable effort to set growth in the city and region on a firm industrial foundation. Pleas for weaning the regional economy away from the "tourist crop" and real estate speculation and reorienting it toward industry were advanced with increasing urgency during the 1920s. A *Times* editorial from 1923, "Balanced Progress," declared that the region stood "at the dawn of a golden tomorrow." But even though the city "glittered" with promise and opportunity, the future was "fraught with great problems" because population growth routinely "staggered all power of anticipation."[30] The editorial identified a set of infrastructure improvements: water and power, an expanded harbor, solutions for traffic congestion, police protection, schools, and parks. In addition, the city needed "men with a large enough vision, prophetic instinct, and practical unselfishness to make these dreams come true." At the same time, entrepreneurs had to reach "further into the back country. To get coal and iron and wool and cotton to feed the industries which will grow, we will be compelled to add great areas of tributary country" (see Figure 9.2).[31]

In response, the Chamber's Industrial Department pursued a proven strategy, cajoling eastern and midwestern industrialists to southern California; it was remarkably successful. A look at one sector, rubber and tire production, provides some sense of the scale and rapidity of change. In 1919, when Goodyear decided to build a Southland plant, only a few independent firms were in production, and these manufactured less than 1 percent of the national output. A decade later, after Firestone, Goodrich, and U.S. Rubber, the nation's number two, three, and four producers, had followed suit, the region's share had grown to 6 percent, which translated into 35,000 tires and 40,000 tubes a day. During that time, employment

FIGURE 9.2. We Must Have a Firm Foundation!
Source: "Balanced Progress," *Los Angeles Times,* 18 November 1923.

went from a few hundred workers to more than 5,000, almost 7 percent of the industry total, and annual output reached a value of $56 million. Los Angeles's location and transit infrastructure were critical factors in terms of raw materials and sales for finished goods. Goodyear's and Firestone's subsidiaries supplied the intermountain and Pacific states, including Alaska and Hawaii.[32]

However, both the Chamber exchange and the *Times'* call for territorial expansion emphasize that the creation of industrial Los Angeles was also a local initiative. As Goodyear and Firestone executives divided up the continent into what economist Frank Kidner labeled "branch plant empires," Los Angeles entrepreneurs and civic elites turned their spatial imaginations to the creation of a "back country" empire.[33] This transition signaled the emergence of Los Angeles as the epicenter of a city-centered region whose entrepreneurs and financiers would exercise influence over dependent territories. It represented enhanced material and symbolic connections in national and international systems of trade and culture as well as an extension of local authority. Like their counterparts in Chicago and New York, Los Angeles entrepreneurs sought to control the hinterlands in two ways: as a center for processing and converting resources and as a center for shaping ideas and marketing culture to shape preferences for consumer goods and exchange.[34]

The distinctions between the east and west sides of the city, between the metropolitan region and other urban centers, and between southern California and other nations were far more complex than these tidy dichotomies suggest. Locally, the Chamber and other pundits minimized the importance of housing and commercial districts for the success of the East Side and other industrial districts. In 1925 the Chicago-based engineering firm Kelker, De Leuw & Co. presented a report to the Los Angeles city council and the county board of supervisors with recommendations for a comprehensive rapid transit plan. This study revealed that, contrary to the received narrative regarding traction, residential dispersion, and sprawl, a significant share of workers employed in the East Side and North Side Industrial Districts lived within walking distance to work, and between one-fifth and one-third of all workers employed in the North Main, East Side, and Vernon districts lived less than two miles from their place of employment. This is the equivalent of the standard distance urbanists accept as a metric for the preindustrial walking city. If the compass is extended to a three-mile circle, it includes more than one-third of the workers in North Main, one-half of those employed in Vernon, and three-fifths of the east side workers.[35]

This type of mixed-use development was the norm for a wedge-shaped segment of the county beginning at the Los Angeles River south of Whittier Boulevard and extending eastward to Montebello and then south along the Rio Hondo to Gage Avenue. This zone encompasses parts of Boyle Heights, East Los Angeles, Commerce, Vernon, and Bell; during the 1920s, it was the site of intensive development. Within these boundaries industrial realtors such as W. H. Daum leased or sold property to B. F. Goodrich, Samson Tyre and Rubber, Union Iron Works, Truscon Steel, O'Keefe and Merritt, Illinois Glass, and Angelus Furniture. Daum began his career as an industrial agent for the Atchison, Topeka, and Santa Fe, opened his own Los Angeles firm in 1913, and over the next four decades helped set the pattern for industrial dispersion in the region. Through a series of holding companies he managed property in the East Side Industrial District while simultaneously developing sections of Vernon and property along Slauson Avenue. In some cases Daum leased land in these new industrial tracts to firms such as the Pacific Coast Planing Company that were moving from parcels he held in the East Side Industrial District.[36]

Daum and his associates in industrial real estate did not simply respond to a perceived opportunity, nor did they meet the needs of a neutral market. Instead, these developers created comparative advantage and developed location attributes. Some elements, such as the rail lines, were already in place. Often the land was sparsely developed, and it sold at attractive prices. But much had to be created, and entrepreneurs like Daum established institutions such as the Eastside Organization and Ninth Street Club to promote collective endeavors. These organizations agitated for street improvements and the spanning of the Los Angeles River with viaducts to "utilize the unsurpassed natural advantages [of the east side]; the facilities of location and uninterrupted territory, the trend of development, and the power of numbers ... to remove all barriers—natural, unnatural, and prejudicial—to its fullest and most permanent development."[37]

Concomitant with this industrial program, other firms, such as the Janss Investment Company, J. B. Ransom Corporation, and Carlin G. Smith, were promoting Belvedere Gardens, Samson Park, Bandini, Montebello Park, and Eastmont (see Figure 9.3). Smith noted that Eastmont, his first subdivision on the east side, was "neighbor to a mighty payroll ... facing a destined city of factories. The amazing development of the great east side—teeming with its expanses of moderate priced homes—has become almost overnight one of the most startling features of the city's growth." The Janss Company, better known for Westwood, Holmby Hills, and other exclusive projects on the city's west side, had been developing

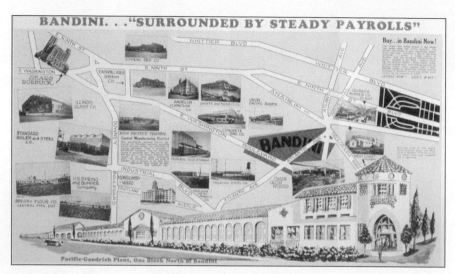

FIGURE 9.3. Bandini ... "Surrounded by Steady Payrolls."

Belvedere Heights and later Belvedere Gardens since 1905; the latter tract, immediately east of the Los Angeles city limits, was intended for "workingmen with limited capital." Advertisements presented Los Angeles as a city awash in humanity "overflowing to the east." The projected exodus to a "new suburb" would bring workers to a district cheek by jowl with "big industrial development." By 1922 the firm was concentrating on parcels adjacent to the Hostetter Tract, where Sears, Roebuck & Company opened a regional distribution center in 1927, and added its voice to calls for widening streets to provide for the anticipated 25,000 new residents "who will make their homes in Belvedere Gardens owing to the great industrial program inaugurated for this section."[38]

During the 1920s, nationally prominent firms such as Swift & Company, Goodyear, Phelps-Dodge, U.S. Steel, Willys-Overland, and Liquid Carbonic established branch plants in Los Angeles. Swift & Company joined local firms such as Deshell Laboratory, Reo Motor, and Sperry Soap in a 300-acre development planned, constructed, and managed by Chicagoans John Spoor, A. G. Leonard, and Halsey Poronto, prominent members of the syndicate responsible for that city's Central Manufacturing District (CMD). These entrepreneurs purchased part of the Arcadia Bandini estate, rancho land that had been held in trust and leased for cattle grazing and farming until 1922, and recast the site for modern industry with large, single-story, fireproof buildings, top-of-the-line services and amenities, and low taxes. Apropos of their Chicago venture, the first phase of development centered on a hundred-acre livestock market and the construction of a central administration building, a terminal warehouse, and a manufacturers' building, with production and storage space for lease to small firms. They subdivided the remaining acreage into 125 parcels, with switch track connections for sale or lease to manufacturing firms. The syndicate offered prospective lessees and buyers financing and construction assistance; infrastructure improvements such as parkways, landscaping, and ornamental street lighting; and the Los Angeles Junction Railway, a beltline with direct connection to all trunk lines entering the city.[39]

Vernon annexed the CMD in 1925, and in 1929 the Chicago syndicate sold out to the Atchison, Topeka and Santa Fe Railroad, which purchased the remaining 2,000 acres in the Bandini estate and extended track and industry west into the remainder of Vernon and east into Commerce. Workers resided in Maywood, Huntington Park, and Bell; the last, "an island of homes in a sea of industry," was an unincorporated community of 9,000 with direct bus service to the CMD.[40] Promotional brochures and publicity photographs for Maywood, the east side residential tracts, and Torrance's model community depict houses under construction on vacant land. Like other frontiers, this crabgrass frontier required an imagined, and in some cases actual, removal of particular groups. Advertisements for the small working-class cottages that took the place of these self-built or "makeshift" quarters made it clear that industrialists and land developers imagined the new east side "miracle city" as a whites-only enclave. In Vernon industrial development proceeded after the burning of informal "shacks" and a "Mexican village." The impetus for this policy of removal and renewal can be traced to an outbreak of bubonic and pneumonic plague. The Los Angeles Department of Public Health and the County Health Department mounted an aggressive quarantine effort, complete with armed deputies. All too easily rat hunting crossed over to witch hunting, and public health officials cordoned off entire districts and scattered house-courts occupied by Mexican nationals and Mexican Americans. Thus the Vernon clearance, an act of city planning, was justified as a public health measure.[41]

By 1930 this configuration of industry, housing, and associated development had become so standard that Thomas Coombs, an engineer with the Los Angeles City Planning Commission, could state simply that "the work shops of the city, the industrial and manufacturing district, should be selected with great care. . . . far enough from the residential section . . . but not so located as to make traveling between the two a disadvantage.

These areas should be large [with] a small part reserved for a local business center. Before it is possible to intelligently subdivide a city, all these subjects should be given careful consideration."[42]

Building a New Metropolitan Region: Oil, Cinema, and Aviation

In almost every municipal and geographic division of Los Angeles, as well as in outlying satellite centers, small industrial districts occur. . . . Hundreds of manufacturing establishments of diverse kinds are widely scattered over the metropolitan district. . . . Even such noteworthy residential suburbs as Hollywood, Beverly Hills, Santa Monica, Pasadena, and Long Beach have their industrial tracts. Airplane manufacturing is rapidly giving a new industrial importance to the Inglewood-Santa Monica district.

—Los Angeles County Regional Planning Commission (1941)[43]

In 1922 the Los Angeles County Board of Supervisors sponsored a conference on regional planning. In the published proceedings a diagram fixes the region as a hierarchy of municipalities, with Inglewood, Hollywood, Pasadena, and Laguna and their spheres of influence arrayed symmetrically around a dominant regional center, the CBD of Los Angeles. The engineers, elected officials, and design professionals in attendance perceived the "whole district crystallizing around natural centers and subcenters . . . each with its own individual character and identity." They envisioned a network of villages connected by transit forming a dispersed but coherent region.[44] If the participants at the Pasadena conference had chosen to extend their study and map the location of firms and plants in the oil, film, and aircraft industries, the diagrams would have coincided to a considerable degree. Firms in these sectors engineered Los Angeles's industrial landscape along the urban periphery during the 1910s and 1920s. This mix of local and national firms produced primarily for export. Petroleum extraction and refining, motion picture production, and aviation engaged relatively large numbers of craft-based workers employed either directly or indirectly in component manufacture and assembly. These firms, in turn, served as generators for the simultaneous and subsequent development of residential districts with associated commerce and institutions.

A dispersed set of "suburban industrial clusters" developed around the oil fields and refinery sites. Refining was remote, but pipelines linked the oil fields to production and tied these dispersed suburban clusters into a coherent network. Standard Oil and other large firms engaged directly in city building, and, in proximity to other production centers such as Brea or San Pedro, smaller-scale entrepreneurs subdivided land and built housing in residential districts such as La Habra and Long Beach. In Vernon, an oil refinery site, zoning and land use regulations restricted residential development, but many oil workers and their families chose to reside next door in Huntington Park and Maywood. Before 1917 four suburban industrial clusters, Whittier-Fullerton, San Pedro–Long Beach, El Segundo–Manhattan Beach, and Vernon–Huntington Park formed a metropolitan industrial district tied by pipelines into a cohesive regional unit.[45]

Although the initial movie colony established an early locus in Hollywood, a 1915 directory of manufacturers recorded a remarkable degree of dispersal, with firms in Long Beach, Santa Monica, Mount Washington, and multiple districts in between. Secondary concentrations also emerged; Thomas Ince established a studio in Harry Culver's new community on the former Rancho Ballona, eight miles west of city hall in 1915. Within five years Goldwyn Pictures, the Henry Lehrman Studios, Sanborn Laboratories, and the Maurice Tourneur Film Company had joined the Ince studio in Culver City, making it the "greatest producer of pictures in the world" after Hollywood.[46] The Ince firm soon metamorphosed into Metro-Goldwyn-Meyer, and in 1927 a promotional pamphlet welcomed Cecil B. DeMille, Hal B. Roach, and the United Artists studios; they had established "plants" in Culver City. The brochure attributed the 71 percent homeownership rate to employment in the studios, "where great forces of men and women are maintained.... As studio development is increased and as the industry becomes more stable, the trend of workers, including professionals, toward the community increases."[47] A similar pattern of development evolved in the North Hollywood area beginning in 1915, when Carl Laemmle converted a former chicken ranch on county land into Universal City. This practice continued into the 1920s with the dispersion of such firms as Warner Brothers, which moved from a site along Hollywood's Sunset Boulevard to an outlying location in Burbank.

Industrial location for aircraft and parts, a critical sector for understanding industry and urban expansion in southern California, fits within this model as well. The origins of southern California's vaunted aircraft (later aerospace) industry can be traced to small, undercapitalized companies that rented space for offices and plants in warehouses and loft buildings in the East Side Industrial District before acquiring more suitable sites along the then urban fringe. In 1912 Glenn L. Martin, who initially directed a crew of mechanics assembling bi-planes in a Methodist church and later a cannery in Santa Ana, founded the first Los Angeles firm to manufacture aircraft, when he moved his mechanics into a brick loft building with a first-floor storefront (a former bedding and upholstery shop) at 943 South Los Angeles.[48]

Donald Douglas, an engineer and Martin vice president, formed the Douglas-Davis Company in 1920, renting the back room of a barbershop at 8817 Pico Boulevard, south of Beverly Hills. Former Martin employees crafted components for a transcontinental plane in a second-floor loft space at Koll Planning Mill, a woodworking shop ten miles away on Colyton Street, near Alameda and Fourth streets. Finished parts were lowered down an elevator shaft and trucked for final assembly to the Goodyear blimp hanger in south central Los Angeles. After securing a contract for three experimental torpedo planes, Douglas, with financial support from Harry Chandler, incorporated as The Douglas Company in July 1921. The following year forty-two employees relocated to a movie studio on Wilshire Boulevard in Santa Monica, chosen for its adjacent field; it proved inadequate for test flights and completed aircraft were towed to Clover Field. Between 1922 and 1928 Douglas produced 375 units, and by 1928 the company had moved its entire operations to Clover Field, which the city of Santa Monica had purchased two years earlier. Municipal ownership assured continuity of operation, the requisite zoning, and eminent domain for expansion. Simultaneously, the firm opened a subsidiary adjacent to Mines Field, an airstrip the city of Los Angeles had recently leased for a municipal airport. By the time the city purchased the property in 1937, the district had become a nucleus for prime airframe contractors, sub-assemblers, and parts and component manufacturers.[49]

During World War II, homebuilders, anticipating an influx of defense workers drawn by these employment centers, selected sites in close proximity for community projects. Adjacent to Mines Field, four developers converted a five-square-mile parcel, owned and planned by Security Bank, into a district for 10,000 residents. A map accompanying advertisements for Westchester in the *Los Angeles Herald-Express* plotted prime contractors and eleven ancillary industries. The copy underscored the district's proximity to a "wide variety of employment." Broadsides enticed potential buyers who could "live within walking distance to scores of production plants."[50]

Westchester, Culver City, Belvedere Gardens, and Torrance cannot be analyzed according to the stock suburban theses. Westchester and similar developments in Los Angeles and elsewhere were not intended as suburbs, if the term is meant to invoke economically inert bedroom communities populated by middle- and upper-income families. Just as significant, these mixed-use districts did not represent a new type of urban landscape, for their antecedents can be traced back at least a century. Finally, Westchester, Huntington Park, the east side, and similar zones throughout the region were highly planned, and, more often than not, they attracted a range of residents in terms of occupation, income, and status, although race and ethnic exclusion remained the norm.

In terms of theory, this Los Angeles case study contributes to a reconsideration of the progressive narratives of industrial geography that begin with technology and sectoral concentration and then follow trains, trucks, and cars out to greenfield sites where industrialists constructed plants designed for beltline operations and mass production. During the 1920s, Los Angeles had a vibrant, mixed-use warehouse, production, and residential district between Alameda Street and the Los Angeles River, organized principally for home-market production, while Chicago-based entrepreneurs developed an industry-only district engineered to the latest standards for mass production firms in Vernon. At the same time, Harry Culver worked with Thomas Ince and other fledgling studio executives to create a film industry satellite along Ballona Creek, midway between the region's central business district (in downtown Los Angeles) and the beach cities. As this study demonstrates, close attention to the particularities of place and circumstance challenges the tidy chronology of ascension, transformation, and succession.

For decades, the Chamber's Industrial Department affixed "Nature's Workshop" to its various publications, advertisements, and pamphlets, but as Raymond Williams, William Cronon, and others have shown, conventional ideas of nature obscure and naturalize relations of power. The notion of climate and sunshine as free externalities purposefully veils the conflict between industrialists and workers and the contest between social reformers, land developers, and city residents, all of whom had a stake in shaping the region. At the same time, the Chamber's sloganeering naturalized industrial development in the southland, and this development took precedence over other considerations. It might be time to reconsider an alternative first posed by the California Commission of Immigration and Housing in 1919. This commission suggested that Angelenos stop being passive recipients of nature's bounty, a "city where nature helps industry most," and become active shapers of a "city where industry helps humanity most."[51]

10

"The American Disease of Growth"

Henry Ford and the Metropolitanization of Detroit, 1920–1940

HEATHER B. BARROW

From 1900 to 1940, Detroit, like all major U.S. cities, underwent a process known as "metropolitanization," in which suburbs experienced more growth than their central cities. This phenomenon peaked in the 1920s, when millions of the nation's jobs and residents moved to the suburbs.[1] Net growth in central cities was, of course, still occurring—the 1920s was the first decade in which the U.S. Census showed a majority of people living in urban places—but the rate of growth in suburbs began to outpace that of cities. Metropolitanization had many causes: cities were overflowing from a growing urban population due to migration within the country, pressure from immigration in previous decades, and natural increase. Meanwhile, people were able to live and work farther apart owing to improvements in the technologies of transportation (the automobile), communication (the telephone), and energy (electricity), all of which were becoming more widespread.[2] Metropolitan settlement was further determined by the dual economic forces of concentration and dispersion: commercial and professional activities were attracted to the downtown, while other functions that required a lower density, such as large-scale manufacturing and residential development, spun out toward the fringes.[3] Suburban

200

migration was also motivated by cultural changes, among them campaigns to "own your own home" and the "Garden City" movements.[4] Finally, metropolitanization increased because of the rise of political opposition to urban annexation. While in previous decades these suburbs would have been incorporated into the cities, beginning in the 1920s they remained separate entities. Even so, they fell under the influence of the central city, economically, socially, and culturally.

Although metropolitanization was common in this era throughout the United States, it happened on a larger scale in Detroit than elsewhere. This can be attributed mostly to the role of Henry Ford, who not only located a factory of unprecedented size on the city's border but also had a greater impact than any other manufacturer in making the automobile a mass consumer item. In other words, metropolitanization in Detroit was spurred by both the migration of the automobile industry and the mobility of automobile users, and Ford was involved with both. Starting in 1919, Ford's relocation to Dearborn caused an exodus of tens of thousands of jobs and residents from the city. It was the second and final step in his pilgrimage from downtown Detroit, following a 1909 relocation of his factory from Piquette Avenue in Detroit to Highland Park. In setting this trend of decentralization, Ford led the way from as early as 1930, as Dearborn was one of Detroit's largest suburbs in terms of population, and it had the highest number of manufacturing jobs, constituting close to a quarter of such jobs in the entire metropolis.[5]

Until this point, Detroit's suburbs had remained small and were of little consequence economically, containing at the turn of the century only 7 percent of the metropolitan area's manufacturing and population, which was one-third of the average for all major metropolises. In this respect Detroit differed from comparable cities like Boston, Pittsburgh, Providence, Cincinnati, and New York, where substantial manufacturing and population had located to industrial satellites during the nineteenth century. Detroit, by contrast, did not experience a significant degree of decentralization until 1910, although once it started, it continued, so that by 1939 more than half of the Detroit-area employment in automobile manufacturing was found outside the city. A year later, in 1940, Detroit's suburbs had the most manufacturing and the second-highest population of any metropolitan area in the United States.[6]

As car manufacturing began to dominate its economy, Detroit was transformed into a city "seething" with "industry and achievement."[7] Automobile plants multiplied across the landscape, as obsolescent facilities were abandoned and newer, bigger ones sprang up in a series of rings progressing from the center, totaling thirty-seven major plants and 250 accessory ones by 1925. Detroit soon turned into a one-industry town. As early as 1914 autoworkers accounted for 40 percent of all factory employees in Detroit, ten times as many as in 1900. As the city attracted more and more job seekers, its population increased from a quarter-million in 1900 to close to a million in 1920, and reached more than 1.5 million by 1940. Compared to other major cities, Detroit by 1920 had a relatively high number of blue-collar workers, foreign-born whites, and men drawn to the city in search of work. In addition, the number of African Americans in Detroit was rising owing to the Great Migration, from 5,741 in 1910 to 41,532 in 1920, or more than four times the growth rate of blacks in Chicago for the same period. The city of Detroit's geographic size grew from twenty-eight to 139 square miles between 1900 and 1925 because of the annexation of suburban territory. Meanwhile, Detroit's rank relative to other U.S. cities was increasing: while in 1900 it had been thirteenth in population and sixteenth in industry, by 1920 it was fourth on both counts.[8]

Even more astonishing was the development of Detroit into a true metropolis, in which rapidly growing suburbs rivaled the central city. From 1910 to 1940 the district outside Detroit expanded almost sevenfold, increasing from 110 square miles to 718, and its population also swelled, notably from 1920 to 1930, when it more than doubled, from 256,588 to 536,102. Because so much growth occurred outside the city proper, the Detroit metropolis as a whole was relatively decentralized by 1940. Growth within the city was also decentralized, since settlement patterns in both parts of the metropolis were determined by a high rate of automobile ownership and a preponderance of one- and two-family homes. In fact, by 1940 Detroit's population density was slightly lower than average compared to other U.S. cities, and this was a notable change from previous decades, particularly 1900–1920, when it was relatively high.[9] Ultimately, the spectacular scope of the Detroit metropolis made it, in the words of Olivier Zunz, "a modern multiple-nucleus city," with "its scores of factories, its center of tertiary activities, its port and railroad terminals, its well-separated production and consumption suburbs, its giant territory, and its diverse population." The Detroit metropolis was, to Zunz, a "symbol of urban-industrial America."[10]

Metropolitanization, however, was never a simple matter of growth at the fringe paralleling growth in the center. It was, rather, a matter of resources being transferred from the city to the suburbs, a process that resulted in losses of various kinds for many U.S. central cities. For Detroit, the burden of these consequences would not be felt until after World War

II, when the number of inhabitants in the central city started to drop in the 1950s and continued to fall in every subsequent decade. Employment in Detroit's central city decreased as well. Between 1950 and 1970 it lost 100,000 automotive jobs alone, while its suburbs gained 40,000.[11] Moreover, Detroit, the country's fourth-largest city since 1920, began to slip behind in national rankings of size. In 1940 it dropped from fourth to fifth place, and it fell again in 1980 and 1990, coming in sixth and seventh, respectively. With the 2000 Census, Detroit gained the dubious distinction of being the first U.S. city, among those with a population of more than a million, to fall below that mark. This drop occurred despite a trend in the 1990s in which most major U.S. cities, even those in the Rustbelt, showed an increase, however modest.

Even during the interwar period, though, discrepancies between Detroit's central city and its suburbs became apparent: the suburbs were more likely to be home to white-collar people than to blue-collar, to skilled industrial workers than to less skilled, to whites than to blacks, to native whites than to immigrants. Meanwhile, the central city witnessed the arrival of a crushing number of job seekers attracted by the promise of factory work; an increase in welfare cases, as wage earners adjusted to the cyclical nature of industrial production; the rise of an inner-city residential slum; economic depression in neighborhoods where industry left; a diversion of industrial wealth from its tax base; and a growing reliance on municipal debt (even during periods of prosperity) for the needed expansion of services and infrastructure. To understand how metropolitanization and its consequences came about in Detroit, it is necessary to examine in depth the motivations of Henry Ford and other automobile manufacturers.

Why Ford Moved to Dearborn

Dearborn offered a number of advantages that made it attractive to Henry Ford as a site for an industrial suburb. Not only was undeveloped land available, but many kinds of transportation served the area. Through the center of Dearborn ran two major rail lines, the Michigan Central and the Pere Marquette, carrying both passengers and freight, an electric interurban line used for commuting purposes, and a highway, now called U.S. Route 12 but known at the time as Michigan Avenue, which connected Detroit and Chicago. Also nearby were other rail lines, such as the Detroit Terminal, the Wabash, the Grand Trunk, and the Detroit, Toledo, and Ironton. In addition, Dearborn featured the River Rouge, which was linked to two Great Lakes via the Detroit River and, once dredged, could be used for shipping. In late 1914 Ford sent his real estate agent, Fred Gregory, to survey land around the Rouge River in nearby Springwells Township for the siting of a new plant, and between 1915 and 1918 he acquired 2,500 acres, close to fifteen times the amount of land he originally acquired for the Highland Park plant.

The huge size of the River Rouge site as well as its accessibility to a heavy-duty transportation infrastructure permitted a degree of vertical integration that had never before existed in industry. The transportation hubs of land, water, and rail developed there would feed the operation with raw materials from the company's own timber tracts, iron mines, and rubber plantation. The complex would also accommodate the processing of raw materials into semifinished products, placing not just foundries but entire steel mills at the location of automobile assembly. Indeed, by attaining this scale of vertical integration, the Ford Motor Company eventually outdistanced all its peer corporations. This was something Ford anticipated as early as 1915, according to company historians Allan Nevins and Frank Hill, who stated, "The project had achieved scope and detail in Ford's mind, for in June of that year he talked to reporters not only of blast furnaces and a tractor plant, but of employing more than 20,000 workers and building a great inland port."[12]

In the meantime, the Dearborn area was relatively undeveloped in 1920, with a minimal level of municipal services and little infrastructure for passenger transportation. In fact, Dearborn's future boundaries were not yet set in 1920. In the area that would become the city of Dearborn lay the township of Springwells and the village of Dearborn, with a sizeable amount of unincorporated land situated between the two. Over the next ten years, Springwells incorporated, first as a village and then as a city, and changed its name to Fordson; Dearborn incorporated as a city in 1927 and a year later annexed Fordson.[13] Throughout the 1920s, Dearborn would gain inhabitants, buildings, businesses, and much more, as Ford Motor Company shifted the center of its automobile production from Highland Park to Dearborn. The city, whose major industries had been limited to farming, brickyards, salt mines, breweries, and a Ford tractor plant, attracted numerous heavy manufacturers, many of which supported the automobile trade. Houses were constructed in the thousands, and a high school, library, and country club were founded. Dozens of streets were paved, and as the roads improved, the number of automobiles increased. In addition, more streetcar lines were created, buses were added to the transit system, and an airport was established. As the population grew, the racial composition changed, and although native whites remained in the majority, the number of foreign-born whites increased, ultimately representing thirty-five different nationalities.[14]

As Dearborn and its neighbor Fordson experienced their rapid and sometimes awkward growth spurt, Henry Ford observed inefficiencies in municipal management that increased the burden for his company and its workers in terms of supporting themselves and their communities. The most obvious problem was that although each of the cities needed to develop public improvements and services at the same time, they had no program for sharing resources that, if implemented, could mean a lower tax rate for both and significant savings for all taxpayers, individuals and businesses alike. Ford, as the largest taxpayer of either community, would benefit the most from this new policy, but his workers would benefit, too. "One central government," he believed, "would be better able to direct the future development of this area, instead of each local community endeavoring to build a city within its own boundaries." To him, this was a matter of modeling municipal management after business management, and in regard to the merger he declared, "The tendency in business today is to lower costs through more efficient methods to eliminate expenditures that are unnecessary—all for the purpose of reducing prices to the public. ... Public funds should be spent in a manner that will serve public need for a greater and longer purpose, which can only be done through centralized administration."[15]

Henry Ford, though, had a few other reasons to push for annexation. He needed a place for his workers to live, blue- and white-collar alike, a large, convenient location able to accommodate thousands of employees within feasible commuting distance. A consolidated Dearborn and Fordson better accomplished this task. Second, Ford needed efficiency in managing his Greater Dearborn holdings, previously situated in three different governmental units: his engineering laboratory was in the city of Dearborn, his estate in Dearborn Township, and his River Rouge plant in the city of Fordson. Dealing with a single government would simplify his financial and legal matters. Finally, Ford needed autonomy in operating his company, a freedom he could be guaranteed only in a strongly defended suburban locale. The logistical challenges of doing business in an urban setting had become apparent to him in running the Highland Park plant. Although Highland Park remained separate from Detroit, it was surrounded by the city in the 1920s, making industrial expansion there impossible. "I don't like to be in the city," Ford stated in regard to the siting of the Rouge. "It pins me in. I want to breathe. I want to get out."[16] Unless they were joined together, Dearborn and Fordson were in danger of annexation by Detroit, or, if not annexed themselves, they could be surrounded by the city, condemned to a fate like Highland Park's. It was thus in Ford's interest to reinforce the independence of a Greater Dearborn, assuring himself room for his company's growth and an arena for his own political power.

Meanwhile, as early as 1920, the dual natures of Dearborn and Fordson were emerging. The village of Dearborn, incorporated in 1893, consisted of 2⅓ square miles lying within the Dearborn Township. The River Rouge plant lay well outside Dearborn Village in neighboring Springwells Township, later incorporated as the city of Fordson. Dearborn had a population of only 2,500 and few industries, functioning mainly as a site of truck farming, with about 400 acres for the cultivation of orchards and vineyards.[17] By 1923, however, the research and development center of the Ford Motor Company was built there, and despite the addition of the Stout Metal Aeroplane Company and Frink Concrete Products in the mid-1920s, Dearborn stabilized as the white-collar twin of industrial Springwells. It was described in a 1926 magazine as a place where "stagnant swamps have become beautiful gardens or willow-bordered pools, magnificent paved highways stretch here and there; cow pastures now hold large apartment buildings; potato fields have become the social center for golf, bridge, and dancing. New churches have been built, a large library, new Masonic temple, beautifully equipped high school, a magnificent country club, [and] a fine athletic field."[18] Dearborn's main street, Michigan Avenue, held banks, stores, a movie theater, the post office, and the town hall, and in the next year Dearborn became incorporated as a city of 9,000. Despite this promising picture, Dearborn soon found itself in debt and, because of its inadequate tax base, unable to keep up with the cost of public improvements. Dearborn had a low ratio of industry to commerce, and since industry was the more lucrative for generating tax dollars, the revenue of the city never reached a high level. Without a proper tax base, it was only a matter of time before Dearborn became financially overextended.[19]

Springwells Township, the home of the River Rouge plant since 1917, was meanwhile going through its own changes. Although Springwells, like Dearborn, had counted farming among its principal economic activities, it was not as much of a stranger to industrial development. Clay manufacture in the township dated back to 1870, and by 1920 Springwells contained the largest brick and tile factories in the state, serving as the primary supplier of bricks to Detroit. In the same year, two metal companies were founded in Springwells, Harwick Stamping and Detroit Seamless

Steel Tubes. In addition, Ford's River Rouge plant was not the only car manufacturer in Springwells; Paige Motor Company, albeit much smaller than Ford, had relocated there in 1919.[20]

As the River Rouge operations expanded, Springwells's citizens responded in 1924 by incorporating the entire township into a city in order to absorb the plant into their tax base and raise revenue, thereby putting themselves in a position to guide the growth spurred by increased industrial activity.[21] They began a public improvements program of several million dollars, creating a disposal plant and a complete sewer and water system. Between 1923 and 1926 new construction starts totaled 3,134 houses and apartments, 65 stores, 81 combined stores and apartments, and 59 industrial buildings, as well as 13 hotels and 2 theaters. These years also saw the construction of 10 schools, 1 bus garage, 1 fire station, and 1 town hall, and the paving of 75 miles of streets. By 1927 Springwells had changed its name to Fordson and grown to a population of 25,000, covering an area of 8.5 square miles with an assessed valuation of $131,250,000. Fordson became a center of industry served by four major railroads, highways, and a port, and host to thirty industries that employed 55,000 workers and created an annual output of $2 million of manufactured products.[22]

Although thriving, especially relative to Dearborn, Fordson was not without its problems. It attracted many less skilled wage earners, as they preferred to live within walking distance of the Rouge plant, and it contained an ethnically diverse neighborhood, home to many immigrants, including Romanians, Italians, Poles, Germans, Russians, Yugoslavians, Syrians, and Turks, to name only a few. As more people moved there, Fordson became a congested, run-down hodgepodge. "Its business, its social activity and its homes," stated resident Iris Becker, "were all mixed together."[23] It was denser than Dearborn, containing, in addition to single-family houses, two-flat, four-flat, and rooming houses, as well as apartment buildings.

Fordson was also more likely than Dearborn to have substandard housing, according to Albert Ammerman, the author of a 1940 sociology master's thesis. Ammerman used as a source a WPA survey that reported slightly more than a thousand families living in substandard housing at rents of $23.50 a month on average. Ammerman referred to the housing as reaching "near slum conditions in many places," with "much doubling up of families." "As many as four families were living in dwelling units made for one family only. . . . In a stretch of four blocks sixty-three houses were found with two families living in a house originally intended for only one."[24] Iris Becker emphasized the connection between uneven housing quality and rapid growth in the South End: "Some of [the housing] was very fast and part of it was self-built. You know, the people who were foreign-born or came from other places built small homes and some of them were larger. They did a lot of work themselves. Some of it was quality work and some of it was not quality work. . . . But many of them, especially the Italians, were good builders." "Those fields just sprouted houses," exclaimed Becker, "Oh yes. They moved right in around on the fields around the Ford Motor Company and walked to work."[25]

When the annexation of Fordson and Dearborn was proposed in 1927, Henry Ford was one of its instigators, and he was primarily responsible for the public relations that caused it to succeed. According to Karmann, mayor of Fordson at the time, Ford approached him and said, "Let's put the two towns together and eliminate the name Fordson. . . . You got a job to sponsor the movement of consolidating these two towns." In return, Ford offered to build a new police headquarters, fire department, hospital, and office space. To win people over, he held a dinner at his estate to which he invited local civic and business leaders. Yet even in this backroom setting, Ford played down his leadership of the annexation movement, saying, "I don't know too much about what's going to happen in the future but whoever suggested this idea, that's a bright and brilliant one." One week before the election, however, Ford unabashedly published a front-page letter in the *Dearborn News* supporting annexation.[26]

Still, despite Henry Ford's support, Dearborn's debt, and Fordson's housing shortage, annexation did not come about easily, and several opposing factions emerged from the population. At stake, after all, was the future of two communities that together had held no more than 5,000 residents twenty-five years before. Opponents of annexation were found mostly in Fordson, which had a lower tax rate than Dearborn and already enjoyed the paved roads, sewage infrastructure, and water main construction that its neighbor lacked. Fordsonites worried that joining forces with Dearborn would force them to subsidize the smaller city's public improvements, thereby increasing their own tax rate. Meanwhile, a faction in Dearborn resisted annexation because it favored smaller government. In the end, a couple of compromises sweetened the deal for both sides. First, the name Dearborn, after the Revolutionary War general, would be used for the consolidated city, and Fordson would be dropped. People were pleased that the Dearborn name was historical, and it happened to by favored by Henry Ford, who was born on its border and considered it his hometown. Second, special assessments on property owners in Dearborn were promised to keep the cost of public improvements from adding to the tax burden of former Fordsonites.[27] Probably the most compelling reason for Fordsonites to consolidate, though, was fear of annexation by Detroit, which would bring about a higher tax rate.

The proposal passed by a three-to-two majority, with support coming overwhelmingly from Dearborn, whose leaders had argued that annexation would boost their city, creating lower tax rates, more efficient transportation, better city planning, improved business conditions, development of the Rouge River as a commercial waterway, and establishment of a municipal hospital.[28] Clyde Ford, the mayor of Dearborn before annexation and a cousin of Henry Ford, was elected the first mayor of the consolidated city. The gains for Fordsonites were much more mixed, however. Although they succeeded in not being annexed by Detroit, consolidation with Dearborn meant an increase in their tax rate—apparently the special assessments on Dearborn never materialized, and Fordson ended up shouldering some of the cost of developing the neighboring community. Henry Ford, though, had many reasons to be pleased, and if the interests of Fordson had not been entirely served by annexation, at least those of his own had been protected.

The Role of the Big Three in Decentralization

To some extent, the decentralization of jobs and residents in the Detroit metropolitan area was a reflection of changes taking place throughout the car industry, which was relocating under the power of the Big Three: Ford, Chrysler, and General Motors (GM). As early as 1930 the centers of production and administration of the Big Three formed a complicated geography that extended throughout the Detroit metropolis and beyond. Of the Big Three, Ford had the most centralized organization, as its production and administrative centers were both located in Dearborn. Chrysler's largest plant, the Dodge Main, was located in Hamtramck, but its administrative headquarters and an additional plant, the former Maxwell one, were in Highland Park. In contrast, General Motors was the most decentralized, with its administrative headquarters in Detroit, small plants scattered in Detroit, Lansing, and Pontiac, and its largest plants situated in Flint. With the exception of GM's administrative center, most of the Big Three's headquarters of administration and production were located outside Detroit, and this was important, for it meant a concentration outside the city, not only of corporate power but also of jobs and residents.

The phenomenon of the "Big Three" was relatively new in the 1930s, and the phrase itself dated back only to 1928, shortly after Chrysler, through its consolidation of Dodge, Maxwell, and Chalmers, became a major competitor of the other two. Previously, in the early 1920s, Ford and General Motors had emerged as the two biggest players in the car industry. Ford had been a formidable presence since the 1910 establishment of the Highland Park plant. A decade later, in 1920, the company's strength was evident in its huge workforce: between Highland Park and Dearborn it employed 57,410 people, or 43 percent of all autoworkers in the Detroit metropolis.[29] Nineteen-nineteen had marked Ford's debut of car production at the River Rouge plant, which by 1930 ranked as the largest manufacturing facility in the world.

General Motors came into being in 1908, when William Durant, head of the Buick Motor Car Company of Flint, purchased Cadillac Motor Car of Detroit, Olds Motor Works of Lansing, Oakland Motor Car Company of Pontiac, several major parts producers, and many other lesser automobile firms. After financial difficulties temporarily forced Durant out of GM, he regained control of the corporation in 1915 and acquired the Hyatt Roller Bearing Company of New Jersey and the Fisher Body Company of Detroit. In 1920 Alfred Sloan took over General Motors, putting into motion the firm's rapid growth, from 12 percent of the automobile market in 1921 to 41 percent in 1933.[30]

When Chrysler arrived on the scene in 1928, its buyout of Dodge, Maxwell, and Chalmers meant that it inherited plants in Hamtramck, Highland Park, and Detroit's east side industrial district near the Detroit Terminal rail line, although Hamtramck contained its largest plant.

Many economic factors that came to the fore in the 1920s contributed to the dominance of the Big Three. First, the absorption of suppliers (and competitors) was increasingly resorted to by the automakers as a way to ensure stability and profitability in an industry that was notoriously cyclical. Second, the growing demands of consumers for model changes incurred significant costs for retooling and advertising, costs that only big firms could afford. As a result of consolidation and competition, the number of active automobile companies fell from approximately 230 in 1908 to 108 in 1923, 44 in 1927, and 12 in 1941. Meanwhile, the market share of motor vehicle sales of the Big Three increased from 40 percent in 1911 to 85 percent in 1937. Owing to this rise, the Big Three by 1939 employed 58 percent of all automotive workers in Detroit, its metropolitan area, and in Flint. In terms of sheer numbers, Ford, Chrysler, and General Motors together employed 128,000 workers that year, the majority of them in plants in Dearborn, Hamtramck, and Flint, communities with a combined population of 264,990.[31]

Residential Decentralization

Both the relocation of the automobile industry and the mobility offered by automobiles contributed to the decentralization of residential development in the Detroit metropolis, and this brought about new configurations of race and class in settlement patterns. Typically, throughout the Detroit metropolis, native-born whites could be found in every kind of community, both urban and suburban. They were usually the majority, although within the city this could vary by neighborhood. By 1940, though, foreign-born whites as well lived in every community, although their presence was the strongest in the city and in industrial blue-collar suburbs such as Dearborn, Highland Park, and Hamtramck. In both the city and these three suburbs, foreign-born whites made up close to a fifth of the population, while elsewhere they were only a seventh. African Americans tended to be segregated, whether de facto or de jure, in Detroit's downtown, but they were gaining visibility in a few suburbs. Overall, blacks still made up 3 percent of the suburban population and 7 percent of the urban, but their presence was increasingly noteworthy. Dearborn is a case in point. There, native-born whites made up most of the population, yet immigrants also had a strong presence, accounting for a fifth. African Americans were forbidden by racially restrictive deeds to live in Dearborn, but many of those who worked for Ford at the Rouge plant moved to the outskirts of Dearborn in another suburb called Inkster.[32]

During the 1920s the population moved outward, even within Detroit, expanding into the "outer city," defined here as the area extending from Grand Boulevard to the city limits. Interestingly, many of the urban pioneers who settled this newly developed district were immigrants and migrants who re-created ethnically homogenous neighborhoods outside the inner city. *Harper's* writer Robert Duffus observed that "newcomers on wheels" caused a retreat of residential exclusiveness, as neighborhoods "gave way with the advance of successive waves of [people] ... who suddenly discovered that it was no longer necessary to live near their jobs or even near the car lines.... Poles, Negroes, Russian Jews, Italians, Belgians, Hungarians, they came in big and little waves, got a foothold, edged their fastidious betters out. Sometimes they encountered resistance and flowed around it, leaving an island to be conquered at leisure or in some cases to be left intact." Regarding the tendency for those of the same ethnicity to live together, Duffus commented that although "a great many Poles live in Hamtramck ... there is nothing but their innate gregariousness to prevent them from moving a few miles north or west where they could get better dwellings and more open space for the same amount of money." "The only generalization one can make about the social map of Detroit," he concluded, "is that lots grow larger and houses more ostentatious as one goes farther out."[33]

While distinctions between neighborhoods were still apparent, overall housing conditions were good in the suburbs and outer city. Detroit, like other midwestern cities, was typified by a low-density landscape of detached single-family houses. Meanwhile, the housing stock was increasing, so that by 1940 Detroit's housing was newer than that in any other metropolitan area save Los Angeles. Rental rates, which had skyrocketed in the teens and early 1920s, declined steadily from 1924 to 1934, approaching the average rents for other large cities, and homeownership increased such that by 1940 Detroit boasted one of the highest homeownership rates (45 percent) of any city in the country. Other conditions were better, too, with lower rates of crime and juvenile delinquency, infant mortality, and deaths from tuberculosis. The rate of welfare dependency was also relatively low.[34]

Despite these gains, residential decentralization meant mobility for only a select number of Detroit's citizens, as many new arrivals, especially those who ranked as unskilled labor, were trapped in the inner city owing to economic or social barriers. Before 1920 many downtown residents were foreign-born whites, and they continued to add to Detroit's population even after immigration restrictions were put in place, for they moved there from other American cities. Chicago, for instance, was an important source of the city's Polish population. The 1920s also witnessed for Detroit the arrival of the Great Migration from the South, which brought great waves of African Americans to the city. As an increasing number of foreign-born whites and African Americans moved into the downtown, taking the place of residents who had moved outward, the city center experienced a severe housing shortage and deteriorated into a slum. Between 1920 and 1940 the inner city remained Detroit's poorest, most densely settled, and most outmoded residential section.[35]

The existence of a slum in the center of Detroit was not entirely new, but its scale was larger than before, both in geographic area and in numbers of inhabitants, and its persistence in the 1920s stood in marked contrast to the prosperity the city was experiencing overall. Detroit's slum was perpetuated, furthermore, by the concentration, both voluntary and involuntary, of the city's newcomers there. On the one hand, the downtown neighborhood offered newcomers of various backgrounds connections to their ethnic and racial communities, which helped them survive by placing them in housing and jobs. In this respect their migration there was to some extent chosen. On the other hand, locating elsewhere was not always practical or even possible. For immigrants, the inability to speak English and the lack of job skills and savings made the downtown the first place to call home until they assimilated and moved elsewhere. African American migrants experienced similar restrictions upon arrival in Detroit, as they too lacked job skills and savings, although they were also faced with a higher level of discrimination. In addition, Detroit in 1920 had some of the highest rents of any city in the country, and the downtown was no exception. Despite its deteriorated housing stock, the slum featured exploitively high rents.[36]

In general, the outward movement of residential and industrial development in the Detroit metropolis meant social as well as physical mobility. When immigrants and migrants were able to escape the inner city, it was because they had the means to buy cars and homes, which in turn gave them new status. Since this trend increased during the interwar years, the ability of newcomers to move outward probably depended on how long they had lived in Detroit. This pairing of outward movement and social mobility caused Zunz to state that, "for the first time, Detroit's spatial arrangement began to resemble the classic zones of successive settlement—which newcomers entered in a set order—described by the Chicago sociologists."[37]

The Decline of Downtown Detroit

Although decentralization brought new opportunities for many Detroiters, it also brought about the decline of the city's once thriving commercial district. At the turn of the century, the downtown served as the center of business and government, with grandiose public spaces such as the Campus Martius, Grand Circus Park, and Cadillac Square. This area also drew and retained the homes of the wealthiest, along with their churches and clubs. The downtown was enhanced in the 1890s by a set of tall, terra-cotta-clad, ornamented buildings in the Chicago School style.[38] Another building boom took place in the 1920s, and according to Thomas Ticknor, "fifteen forty-five story towers such as the General Motors Building, Book (new Detroit)–Cadillac Hotel, Buhl Building, Fisher Building, Penobscot Building, and Guardian Building recast the local skyline."[39] Cyril Arthur Player referred to Detroit's skyscrapers as "newly American in architectural form and fit to grace a city of ultimate beauty."[40] These buildings gave the downtown "the air of a metropolis," commented *Munsey's* writer Judson Welliver, who added, "Skyscrapers are everywhere, magnificent shops occupy palatial quarters, and the idea is that whatever is good enough for Detroit must be a little better than anything else in its class."[41]

The optimism and energy typical of the downtown during this era drew the attention of many observers. According to a 1915 writer for *Outlook*,

Every one in the Michigan metropolis is buoyant, enthusiastic, with an engaging point of view.... The crowds move along swiftly, eagerly, but without that tension which is characteristic of a mass of New Yorkers.... There you have the spirit of the place—the spirit of do and dare to accomplish things, the intense desire for activity, the penetrating belief in the great destiny of Detroit, and the supreme joy of achievement.... Since the days are not long enough in Detroit, this bounding, effervescent city must needs labor into the night. Daylight reveals stately structures towering, in all their unblemished newness, above the streets that with each passing month become more crowded and more cosmopolitan in character. Until well into the evening the shops and stores still hold out their lure to the throng of buyers, and the great buildings are illuminated shafts against the sky.[42]

All the same, development of Detroit's downtown was modest compared to other major cities, according to Ticknor, who notes that "although Detroit's sixteen million square feet of office construction in the twenties ranked fifth among American cities, comparison with other cities indicates the relative weakness of the Detroit office boom. Between 1919 and 1929 New York gained seven times and Chicago 2.3 times more office space than Detroit, and Detroit's office construction per capita ranked last among major cities."[43] As commerce played a relatively weak role in Detroit's economy compared to manufacturing, the development of office buildings was limited, and the downtown that contained them was relatively small.

This situation was exacerbated by GM's decision to develop the New Center, a group of office buildings, including the company's administrative headquarters, at a location within the city limits but about 2.5 miles from the downtown. In addition, a site near GM's New Center was developed as the Center for Arts and Letters, which consisted of the public library (1921) and the Detroit Institute of Art (1927). Also adjacent was the existing campus of Wayne University. Once realized, this group of cultural institutions combined with the commercial development of General Motors to form a rival downtown, one complete with art, education, business, retail, and entertainment.[44] Meanwhile, the choice of Ford and Chrysler to maintain their administrative headquarters in Dearborn and Highland Park, respectively, further weakened the downtown.

The working-class culture of Detroit also made the downtown less relevant as an attraction for shopping and entertainment. "The crowds of pedestrians downtown ... soon thin out," observed WPA writers, who claimed that the "most exciting spectacle" was to be near one of the large factories at the end of a shift: "Shrieking whistles at the end of the day signal the end of the work period, and the factory disgorges a veritable flood that fills the streets almost from curb to curb. It is a flood, not of men solely, but of automobiles, and on the steering wheel of each are the calloused hands of a workingman." Detroit may have "all the other externals of the metropolis," said the authors of the WPA guide, "but it lacks something of the bloom and glitter of such cities as New York or Chicago," and they attributed this to the fact that "Detroiters work hard. The bulk of them have little time for culture, for the theater, the night club, or the erudite lecture. They find their recreation in going on Sunday drives with the family or cheering their favorites at the baseball park. 'Doing the night spots' consists mainly of making the rounds of beer gardens, burlesque shows, and all-night movie houses. Only one legitimate theater manages to survive the arduous Detroit winter."[45]

The onset of the Great Depression only added to the problems of Detroit's downtown. According to Ticknor, "Vacancy rates, demolition of buildings to create parking lots, and building obsolescence increased substantially in the downtown area during the thirties, and income from rentals in the central business district fell by 50 per cent between 1929 and 1940."[46] Nearly a hundred commercial buildings were torn down because the owners could not afford the taxes.[47] In their place, parking lots (referred to as "taxpayers") were operated, as they made more money than leasing office or retail space. In 1940 local historian Arthur Pound complained of the "ugliness of the average parking lot and of the back-walls, never built for public gaze, that surround it," calling such structures "scars [that] cloy the near view of business Detroit." "The use of downtown land for ground-level parking only," he continued, "is neither thrifty nor socially defensible in normal times."[48] As the downtown deteriorated, wealthy residents left it for exclusive suburbs, with the result that their Georgian mansions were carved into rooming houses. Soon the spread of car showrooms, gas stations, and storefront businesses up Woodward Avenue disfigured a fashionable residential district north of the downtown.[49]

Even worse, the downtown failed to recover after the Depression ended. Growth had come to a "practical standstill," according to one Detroit traffic engineer. Less than a quarter of the metropolitan population entered the downtown every day, a drop from its peak in the late 1920s. "Detroit, like other major cities of the nation, is faced with the problem of decay at its heart," commented the city's planners. "To the initiated in the problems of a metropolitan area, the story of blight needs no explanation. All are familiar with basic characteristics: the decline of property values, the flight of old residents, the increased ratio between tenants and home owners, and the growth of slums."[50]

The role of the automobile must be considered as part of the explanation for why the downtown remained deteriorated even after the return of prosperity. "Detroit has been hurt as well as helped by its own mobile creations," stated Pound, adding:

> Around the business section extends a dismal zone of neglected structures, many of them once the proud habitations of leading families, whose owners now live in the outlying belt. Their former dwellings, put to nondescript uses, form an unsightly barrier between attractive business and residence areas. Further out, in newer developments accessible by motor-car, one finds domestic architecture and landscaping at a high general level even in modest neighborhoods, and rising toward impressive heights of beauty and dignity among the opulent.[51]

"In Detroit, little money seems to be available for investment in and improvement of properties in decaying regions," he concluded. "The chief reason is ease of movement—it is as easy to get to work to-day from twenty miles out as it used to be from four miles out. . . . Where daily commuting and shifts in location are undeterred by natural obstacles, as in level Detroit, there is little incentive to rescue those decaying neighborhoods that the townsfolk resignedly call 'blighted areas.'"[52]

During the interwar period, according to Dearborn mayor Clyde Ford, the rise of the automobile industry wrought in Detroit and its metropolis a change of "marvelous proportions." Mayor Ford believed that the industry had led to a "gradual awakening" that "caused the shadow of expansion to spread like a great magic force over the surrounding territory." He continued, "More changes have come to this surrounding territory of Detroit than would come to most communities in a century." In his opinion, this was "caused and started by the automobile," and "the very air has been permeated with the speed of the whole thing. Production and efficiency are words common to everybody, and living in this area of progress and change . . . has caused the name of 'dynamic' to be properly applied to the city of Detroit." "All this," he mused, "had its start with the building of the first little Ford car by Henry Ford."[53] "In a single decade sprang up the colossal factories that today produce a third of all the automobiles, trucks and tractors in the entire world," observed writer Thomas Munger. "Smoke from a thousand chimneys darkened the skies, and in a thousand plants the roar of machinery sounded."[54] Signs of automobile manufacturing and the labor it entailed spread throughout the city, which was marked by "staring rows of ghostly blue factory windows at night, the tired faces of auto workers lighted up by simultaneous flares of match light at the end of the evening shift; and the long, double-decker trucks carrying auto bodies and chassis."[55]

Many remarked upon the enormous influx of people into the Detroit metropolis. *American Magazine* writer H. M. Nimmo commented, "Men who have heard of our millennial wage scales have hit the trail for Detroit and dreamed dreams of Eldorado. Farmers grown restless in their bucolic surroundings have shaved off their whiskers and joined the procession. Venerable hoboes make a Mecca here, and enterprising crooks in search of easy money lavish their forbidden attentions upon us." Robert Duffus observed in *Harper's* that although a *Detroit News* guide to the city had described the locals as "a joyous folk," he himself witnessed "little that is joyous about Mr. Ford's factories or any of the other factories in and around Detroit which have been forced to imitate his methods. One hears of crowds of workingmen coming out at quitting time too drunk with weariness to talk." Munger noted that "the population multiplied, and in the shifting crowds along the streets appeared the alien faces of laborers from all the congested cities of Europe."[56]

Despite its ascent as one of America's leading cities, Detroit lacked cosmopolitan flair. "There is atmosphere, yes," wrote the *Century's* Webb Waldron, "but its single characteristic is a smell of gasolene [*sic*]." "Imagine this," he continued, "a cluster of new sky-scrapers thrusting gawkily up out of a welter of nondescript old buildings. A big open square crowded with automobiles; great radiating streets teeming with crowded trolley-cars, radiating outward like the fingers of a Brobdingnagian hand through a vast dreary waste of criss-cross streets lined with rows of soot-blackened wooden houses, on and on, mile after mile, till they reach stupendous palaces of steel and glass that suck in and disgorge hundreds of thousands of workmen morning and night." Duffus agreed, writing that "Detroit . . . is not a majestic or beautiful city. . . . Certain imposing views there indeed are. . . . But not for a single moment does it make one tingle." Meanwhile, Edmund Wilson, writing in *Scribner's*, referred to the New Center built by General Motors as a "bulky herd of thick, square, Middle-Western skyscrapers."[57]

The descriptions of Detroit's suburbs, which emphasized their affluence and beauty, stood in contrast to those of the city. "On all sides are reflected the many activities that reach far out into the suburbs," stated a writer for *Outlook*. "There are great stretches of residential sections and streets and boulevards that in many cases were open country a year ago. Chicago and

Boston Boulevards, two of the fashionable thoroughfares of to-day, flanked by imposing residences, are little more than a year old. Along Lake Saint Clair, property that a few years ago sold for sixty-six dollars a foot now brings in one thousand dollars." *Munsey's* writer Judson Welliver even compared some of the grand houses in Detroit's suburbs to those in Newport, stating, "Palatial 'cottages' in wonderful grounds look out over the blue waters of the lake, miles of them lining the shore."[58]

This disparity between Detroit's city and suburbs can be attributed to a lack of planning during its spurt of metropolitan growth, one that was criticized at the time. Duffus, for instance, complained of the decentralization of urban development, writing that Detroit "has been able to spread itself over an almost limitless area, with few natural obstacles" because of the "almost universal adoption of the automobile and the frantic use thereof." Detroit suffered from the "American disease of growth," he concluded.

A letter to the editor of the *New Republic* blamed the automobile manufacturers, who, she wrote, "have been content to see the city lose much of its charm and seemliness; trees are gone, ugliness is everywhere in a mushroom growth of cheap buildings; dirt and untidiness are the despair of an overworked board of health, politics are well meaning but impotent, and a tolerable climate has been murked with a smoke pall that rivals Pittsburgh." This writer asserted that while at one time, "in the days before the Selden patent, Detroit was a beautiful city—of clean, wide streets, luxuriantly shaded, of spacious dignified homes, of tolerable climate," these days, "anything goes that does not interfere with production." Thomas Munger commented on the impossibility of getting away from pollution, as the factories were so entangled with the city's fabric. "The rapid increase in population since 1910 has caused the city to spread out, so the many large plants, which only a few years ago were located far outside the city limits, now are found in the center of some otherwise pleasant residence district, or even far downtown in the neighborhood of office buildings, shops and theaters. . . . Residence property values in many sections of the city," he concluded, "have been seriously affected by the smoke and noise of the factories."[59]

Detroit, moreover, had not gone through this period of drastic growth without some attempts at planning its development. As early as 1919 *Detroit Saturday Night* writer Harvey Whipple complained, "Detroit's growth has been like Topsy's; it has been without plan; it has *happened*; it has *occurred* as the immediate occasion dictated and individual development has always been paramount to the general public good. . . . There has been no scheme by which future growth could be directed and controlled; no plan by which investors could be reasonably sure of the values of their property. A valuable piece of real estate today might be utterly worthless tomorrow." Whipple called for the creation of a zoning ordinance and noted, "other cities have been through this struggle of heterogeneous and undirected growth and development.... It is not so long since that the problem was solved in New York city only after great damage had been done and many personal losses experienced."[60]

In addition to zoning, there were demands for metropolitan planning, and one advocate of this was Frank Burton, Detroit commissioner of building and safety engineering. Noting that a "survey of construction throughout the Greater Detroit district shows single home building forging far ahead," he asserted: "There is every indication that this building expansion will continue until the city runs into all the adjacent counties.... The advent of such a condition brings new civic problems that must be provided for in a really big way, and some form of incorporation [of a city-county improvements authority] must be adopted if policing and sanitary features are to be controlled."[61] Although a number of similar plans were proposed throughout the 1920s, there is no evidence that they were implemented in more than a piecemeal fashion.[62]

Overall, the metropolitanization of Detroit was undertaken without much foresight, spurred as it was by industrialists spreading their factories around the city's periphery and by residents who pioneered the "crabgrass frontier" in their automobiles. Ultimately, Detroit could do little to bridle the growth of its metropolitan area without the cooperation of the Big Three. The automakers, however, had their own agenda, and by 1940 Ford, Chrysler, and GM all maintained their major centers of production outside the city. Two of the three also located their administrative centers outside the city, and even GM, which had its headquarters in Detroit, chose for it a site away from the downtown. For the automakers, Dearborn and suburbs like it offered ever cheaper land and labor the farther away from the city they were situated. At the same time, increased promotion by local and federal officials of automobile use (through road-building) and homeownership (through mortgage standardization) meant the automakers could count on workers coming to them to live and work. Finally, sponsoring suburbanization, both industrial and residential, was in the interest of the automakers because it increased dependency on their own product. Furthermore, to Henry Ford at least, suburbia appeared to offer a better life for his workers, something in which he, as a welfare capitalist, took an interest.

The result of all this was the metropolitanization of Detroit during the interwar years, a process by which the rate of growth in the suburbs outpaced that of the city, although the city population was still growing. In the meantime, the rise of industrial suburbs like Dearborn worked a devastating effect upon Detroit, weakening its commercial center, increasing slum and blight, hardening the geographical boundaries of race and class, reducing its number of jobs, and shrinking the tax base.[63] Although this situation culminated in the Great Depression, when the city found it no longer had the resources to support its citizens, the full ramifications were not to be felt until after World War II, when the city population began an actual decrease.

11

Suburbanization and the Employment Linkage

RICHARD HARRIS

When North Americans speak about suburbs they usually mean the sorts of low-density residential environments in which most of us now live. These are the sorts of landscapes that have attracted the attention of urban historians.[1] For many years scholars documented the genealogy of the affluent suburb, showing how it was adapted and diluted to form the modern planned subdivision. Recently, some have attended to the grittier places that were settled by workers and immigrants, including the neighborhoods that grew up beside suburban industry. Even here, however, the factory chimney lies in the background of the picture, offering plausible depth to a scene that is still largely domestic.[2] More commonly still, it has been airbrushed out.

The essays collected here are written from a different standpoint, one that puts the factory—and, if only implicitly, the office functions that often went with it—in the forefront.[3] The view is invigorating, if sometimes grim, and brings us closer to the balanced understanding to which scholars usually aspire. To understand suburbanization, however, we must do more than add industry to the mix and stir. Several contributors suggest that the suburbanization of homes depended on the decentralization of industry, and some hint that the reverse was also true. These were linked processes. This chapter explores that important but neglected truism.

221

The connection of work and home—what James Vance has termed the employment linkage—is vital to the changing form of urban areas.[4] It has been neglected because, as is often the case, we have allowed available evidence to shape our research. There is abundant information, for example in the decennial census, about the location of housing and of factories, and this tells us about the character of specific suburbs, and of suburbs in general. Scholars have applied descriptive terms such as the "residential" or "industrial" suburb to describe places with, respectively, a surplus of people or of manufacturing jobs, along with an intermediate category for "mixed" or "balanced" suburbs. Such terms are applied to places that have constituted themselves as municipalities but that often lack a functional identity. They describe the results of suburbanization, not the process itself. The existence of a mature industrial suburb, for example, has often been taken to indicate that industry led the way into the urban fringe. In fact, such a suburb could have evolved in different ways, beginning as a mixed or even a residential community, before acquiring its industrial character. Its mature form is a fallible guide to the formative process, and the same is true for other types of suburbs. I argue that to understand the long-term spreading out of homes and industry, we need to worry less about types of suburbs and think more about the process by which they came into being. In the first section of this chapter I elaborate on this point, distinguishing between the processes of residential, industrial, and balanced suburbanization, as well as a compound type that I call alternating development.

The different processes of suburbanization have been active in every urban area, but not to the same degree. Scholars, including many of the contributors to the present collection, have suggested why suburbanization took particular forms in specific urban areas, but in an *ad hoc* manner that has not led to systematic reflection. For this reason we have little idea whether the importance of the different processes changed over time. In the second and third sections of this chapter I discuss the causes of local variation before considering the question of historical trends. The arguments that I develop are speculative. To test them it will often be necessary to undertake time-consuming research with primary records. In the final section I indicate how such research might proceed and why it should be undertaken.

Processes of Suburbanization

Logically, the suburban trend can be led by jobs, by residential settlement, or by a mixture of the two. Until the third quarter of the twentieth century the most common types of jobs were in manufacturing.[5] This suggests three processes of suburbanization: industrial, residential, and balanced, with the last involving the joint decentralization of roughly equal numbers of workers and jobs. It is easy to find examples of places that were produced by one or another of these processes. As Richard Walker describes it in Chapter 6, Emeryville, California, was driven by industrial development; Mary Beth Pudup shows that the commuter suburb of Riverside, west of Chicago, was first and foremost a residential project, and that Harvey, Illinois, was designed to contain a mix of homes and industry.[6] Each place retained its original character over a long period. But historical development is rarely linear. To be successful, places planned for industry soon needed to acquire homes. In so doing, they attracted other manufacturers, who in turn drew workers whose settlement reinforced the suburb's appeal to employers. A similar recursive logic applied to many residential suburbs, especially those of workers. Viewed historically, then, what I term "alternating development" constitutes a fourth type of suburbanization, one that combined the other three types in a great variety of ways.[7]

We have only the vaguest idea about the relative importance of these four processes. We cannot infer their prevalence from their results. As noted, the mature suburbs with factories and workers' homes that are usually labeled "industrial" could have evolved through processes of industrial, mixed, or alternating development. They might even have begun as residential enclaves. Norwood, in the Mill Creek Valley just north of Cincinnati, is a case in point.[8] In about 1914 Norwood was a booming industrial area with a shortage of housing. Workers were compelled to commute out from Cincinnati. It might seem that here was the muscular beginning of a classic industrial suburb. But Norwood, and the adjacent suburb of Oakley, had in fact started as "residential suburbs of the usual type."[9] These settlements, and the labor force they contained, probably helped to attract industry to the area. Without doubt, the final character of the area did not reflect its beginnings. Such indeterminacy would routinely have been true of what ended up as mixed suburbs: these could have evolved in a wide variety of ways.

It might seem that wholly residential districts such as Riverside imply a clearer and a simpler story. Often this was the case. Even here, however, we must be wary of ambiguities and shifts. Many places that had the reputation for being residential in character were in fact more mixed. An excellent example is Parkdale, a "leafy suburb" of the late nineteenth century that at first had the reputation of being one of Toronto's premier residential suburbs. In fact the public image, promoted by developers and local boosters, disguised the fact that the area always contained some industry, as well as workers' homes.[10] Then again, a purely residential suburb may be just as accessible to workplaces in an adjacent suburb as it is to employment downtown. To the extent that it was first settled by downtown commuters, it is a product of residential suburbanization; to the extent that its residents worked in a neighboring industrial suburb, it should be regarded as part of a process of alternating suburbanization. Many "residential suburbs" blended these two processes. Earlscourt, a working-class suburb of Toronto, is a case in point. At first, many residents commuted downtown, but a large minority found employment in a nearby industrial district.[11] In such situations we cannot infer the process at work from the residential character of the district in question. Even its location in relation to centers of employment is only a rough guide to the forces that made it.

In disentangling these forces we should pay some attention to municipal boundaries, but not too much. Municipal boundaries often influenced the process of suburbanization: suburbs sometimes gave tax breaks that encouraged factories to locate there; weaker building regulations in fringe areas often drew workers who might otherwise have remained within the city. But we should not allow political boundaries to define the process. In cities that contained vacant land, suburbanization could occur within city limits. Very commonly it straddled those limits, even as they evolved. Flint, Michigan, offers a good example. In 1916 a residential subdivision was created in Atherton Park, south of the city limits.[12] Residential settlement began immediately, but less than 14 percent of the area was developed before it was annexed in 1920. Four years later an auto plant was built nearby, just within the city, and this encouraged settlement both within and beyond city limits. By 1945 more than three-fifths of the lots in Atherton Park were still vacant, though many were built upon over the next six years. Here, then, an "alternating" process of suburbanization straddled city limits over a period of decades.

Although we do not know the relative importance of the different types of suburbanization, all have been present in every major metropolitan area in North America. Clearly, however, the balance between them has varied. In this volume, Richard Walker and Edward Muller have argued, respectively, that industrial suburbanization shaped metropolitan San Francisco/Oakland and Pittsburgh. It was also, in Heather Barrow's account, a force in Detroit. Elsewhere the residential dynamic was preeminent. New York City is the largest example. There the suburban settlement of people proceeded earlier and faster than the decentralization of offices or manufacturing did.[13] An obvious explanation is the early development of a cheap and efficient transit system, but other, smaller centers that lacked suburban transit have also been dominated by the residential type of suburbanization.

Although it offers examples of alternating development, Flint is a fine example of residential suburbanization, illustrating the emerging significance of the automobile from the 1920s. Flint was itself a major center of the auto industry, but until 1947 all the major assembly and parts plants remained within city limits. Even so, by 1940 a quarter of the people in the metro area lived in suburban fringe areas.[14] These suburban settlers were motivated overwhelmingly by a desire to own homes: a 1948 survey found that this was the main motive of one-third of those who had moved to the suburbs in the previous decade or so.[15] Industrial decentralization had played little role: the same survey found that barely 3 percent had moved to the fringe in order to be closer to work.[16] By the 1930s the car was making residential suburbanization very possible. In 1936 two-thirds of those who were employed at four of the larger auto plants used cars to get to work, a third as passengers.[17] (Car pooling, often advocated but rarely practiced in recent years, was once common.) The opening of a new Chevrolet plant in the suburbs in 1947, then, added weight to a suburban trend that was already well under way.[18]

In terms of the forces that shaped suburban development, Pittsburgh and Flint fall toward the extremes. In most metropolitan areas, the suburban dynamic was more balanced. During the nineteenth century, a good example was Philadelphia, where residential and industrial suburbanization moved in step.[19] This type of urban growth could reflect either of two possibilities. In some metropolitan areas, balanced growth in the suburbs as a whole was the result of the balanced development of individual suburbs, or at least of each suburban sector. According to Greg Hise, Los Angeles provides an example where, as early as the 1920s, land developers promoted the linked development of homes and industry in districts such as Leimert Park. This goal was then elaborated on a regional scale in the 1940s in places like Panorama City. Alternatively, and more commonly, different forces were dominant in different suburbs, so that it was only the overall metropolitan experience that might be described as balanced. One such example, described by Gunter Gad, was Toronto. There, the suburb of West Toronto Junction and the satellite of New Toronto grew up around industry, while other fringe areas were largely residential. In aggregate, these evened out. Robert Lewis's survey suggests that the same mixture was apparent in Montreal, which included industry-led suburbs such as Lachine, residential suburbs such as Mont Royal, and the planning of a mixture of homes and industry in Maisonneuve.[20] Another example was Cleveland, with its combination of residential suburbs like Shaker Heights, industrial suburbs like East Cleveland, and balanced suburbs like Lakewood. Significantly, a survey of household mobility undertaken in the 1940s found that the desire to be closer to work played almost no role among those who settled in Shaker Heights (5 percent), but that it figured prominently among those who had moved to East Cleveland (19 percent) and Lakewood (21 percent).[21]

In all of these places, alternating development was the norm. Although none of the contributors to this volume seeks to document it systematically, many hint at its importance. Lewis, in particular, suggests that in Montreal the success of industrial districts routinely depended upon their later ability to attract workers. He has also indicated that some industrial suburbs like Lachine had grown from a small, preindustrial base.[22] Similarly, for nineteenth-century Chicago, Mary Beth Pudup indicates that Brighton Park began as Brighton, a small trading settlement that grew because its local labor force was able to attract industries and then because those industries in turn attracted workers. Recursive development was the norm in industrial districts that contained many small firms. Part of the logic of such districts was the creation of a shared, skilled labor force, and as this labor pool was built up the district gained momentum. The same back-and-forth growth of housing and industry was less apparent but could be no less real where one or two large employers ramped up production in a series of steps, as the local labor supply allowed. It could even play a minor role where large operations were established overnight. Around Detroit, for example, Barrow emphasizes that Ford's huge investments at River Rouge and Highland Park gave a major stimulus to suburban settlement, but she also suggests that the location of these factories had been encouraged by the prior availability of at least some labor nearby. Alternating development added an element of dynamic complexity to every other type of suburbanization, and in every metropolitan area.

The only places in which balanced development was mandatory were new satellite communities. In the early twentieth century, contemporaries distinguished industrial "satellites" from "suburbs."[23] The former were distant, and distinct, from the main metropolitan center. Almost by definition, they drew few workers from beyond the limits of the community. Here industry—or the promoters of new industrial development—had to take the initiative, otherwise households would have no reason to settle there.[24] Gary, Indiana, was the largest of this type. At the same time, no company would invest in such a place unless housing was available from the outset, even if this meant building for their own workers. In Gary, U.S. Steel built company housing, chiefly for skilled workers, and helped ensure that developers, builders, or workers themselves would meet the remaining needs of the labor force.[25] In the Pittsburgh region a number of industrial satellites were established. Muller indicates that in every case employers satisfied themselves that homes were available, sometimes but not always by building company housing.[26] Satellites, then, had to contain a mixture of homes and industry from the very beginning. In contrast, because industrial suburbs were extensions of the larger center, they could develop in more diverse ways. The line between satellite and suburb was fuzzy, and it shifted as metropolitan areas expanded and as new, cheaper methods of transportation became available, but the distinction remained meaningful.

Reasons for Local Variation

Historians have offered a variety of reasons as to why, in particular situations, one type of suburban dynamic was dominant. The role of the subways in New York, for example, and the impact of the steel industry on the Pittsburgh region are well appreciated. How may we systematize these observations? Singly or in combination, five linked factors determined whether industrial or residential development led the suburban trend: the nature of local industry; the stability of the relation between employer and employees; the character of the labor force; the state of the local and national labor market, and the availability of mass transportation.

An obvious influence on the nature of suburbanization is the nature of local industry. Certain industries were drawn and remained tied to central locations. Garments and printing are the leading examples. In cities such as New York, where these industries were prominent, industry was less likely to be a dominant suburbanizing force. Other industries, notably steel, meatpacking, and auto assembly, favored locations at or even beyond the urban fringe. The results may be seen in Homestead, Pennsylvania, in Chicago's Back of the Yards, or in Hamtramck, Michigan.[27] Of course, patterns of industry were not wholly predictable. As already noted, for example, in Flint the auto plants were slow to decentralize. But in general the city's industrial structure was a major determinant of the pattern of suburbanization.

A less obvious, and less well understood, factor is the character of the relationship between worker and employer. Companies that had invested in training a loyal workforce might have been reluctant to relocate for fear of losing this resource. If it had won such loyalty by offering secure, well-paid employment, however, a company might expect that its workers would soon follow, especially because, in moving to the suburbs, they would gain access to more space and cheaper housing. More attention has been given to the converse possibility, where employers were eager to shed an existing, troublesome workforce and where they viewed relocation as an opportunity to fashion a more pliant workforce. It is unclear how common this consideration might have been. Gad suggests that, at least for a couple of decades, in Toronto it was the suburban worker who was the more militant. Companies were likely to be most willing to abandon an existing labor force that had limited, and easily replaceable, skills, as on the assembly line at Highland Park.

A countervailing influence could arise where workers were employed on a casual basis. Lacking stable work, laborers preferred to live where they had ready access to a range of potential employers. Historically, this would commonly have been a central location, around docks and warehouses. Those industries that relied on a casual labor force had an incentive to cluster, since this limited the pressure to provide job security. Such industries, subject to the vicissitudes of weather or of unpredictable demand, included the fashion-oriented segments of the garment industry. Here, neither employers nor workers were keen to suburbanize. The only exception would occur where several firms relocated simultaneously and in large numbers, as happened with the almost instantaneous growth of packing plants around the Chicago stockyards.

The significance of the employment relation was complicated by the character of the labor force, especially its gender composition. To this day in many families, women's work is seen as a supplement to household income. In most households women have earned less than men and have shouldered most domestic responsibilities. Both considerations discouraged them from commuting long distances. This pattern was confirmed when families acquired their first automobile, since this was almost invariably used by the man of the household. Among Flint's autoworkers in the 1930s there was a significant number of women, especially at AC's two sparkplug factories. In 1936 almost all of the women who were employed at the AC plant on Industrial Avenue walked or took transit. At three other auto factories, assembly plants for Buick, Chevrolet, and Fisher Body, the great majority of employees were men and about three-quarters commuted by car.[28] The lesser social and financial status of women's work meant that households rarely moved to follow the wife's job. For all of these reasons, employers and industries that relied on women could not expect to retain their labor force if they relocated to the suburbs. For example, in the early 1900s, when some garment manufacturers moved from Lower Manhattan into the Boroughs, or even uptown, their female labor force did not follow.[29] The businesses in question had to tap a new suburban labor force. Since they must have been aware of this fact in advance, the owners were probably careful to relocate into established districts where there was already abundant labor nearby. Another example, which makes explicit the differential impact of industrial decentralization on men and women, concerns the relocation in 1917 of Kodak's Canadian head office and camera assembly plant from a fairly central location to Toronto's fringe. A comparison of its labor force before and after the move shows that more men than women stayed with the company, and many followed it into the suburbs.[30] The home-work dynamic, then, was gendered in complex ways. Industries that employed women and that valued their current labor force would have felt pressure not to suburbanize. Those that employed men whom they deemed dispensable would have felt few inhibitions about relocating. Because such considerations varied by industry, and also by company, they shaped a unique balance of suburbanization in each metropolitan area.

A fourth and neglected influence on suburbanization is the tightness and stability of the labor market. In periods, or regions, of labor shortage employers are wary of moving to fringe areas that have a limited labor supply. At the very least they need to make sure that adequate housing will be available. We might hypothesize, then, that during economic booms balanced development is likely to be quite common. In contrast, during periods (or in areas) of high unemployment, employers may be more inclined to assume that if they build the factories then workers will follow. Economic downturns may have favored industry-led growth, although there were not many examples of this, since in such periods development of any kind is usually slow. It is not difficult to find prominent examples that fit these speculations. Pullman, for example, was planned as a balanced suburb during the boom years of the 1880s. The thinking of George Pullman went beyond the usual concerns, but it was recognizably shaped by the challenge of attracting Chicago-area workers into a largely rural area at a time of relative prosperity.[31] In contrast, when Caterpillar expanded operations in East Peoria, Illinois, after 1932—unusual timing that depended on the company's ability to secure heavy equipment contracts from the U.S. government and from Soviet collective farms—it gave no thought to the housing needs of its new labor force.[32] At any other time this would have been remarkable, for the company could not even rely on local developers and builders to play their part: in 1932–33 the building industry in the Peoria area was virtually moribund. If the population of the adjacent suburb of Creve Coeur boomed overnight, then, it was because workers, including migrants from the coalfields of southern Illinois, were willing to build their own homes.[33] A similar pattern of suburban development attended the construction of Ford's mammoth bomber plant at Willow Run, near Ypsilanti, Michigan, in the early 1940s. This project was planned with a reckless disregard for workers' needs.[34] To some extent the exigencies of wartime help to explain this neglect, but the precise timing of the project, coming as it did at the tail end of a decade of an unprecedented labor surplus, was surely also important. Examples such as these illustrate, and allow us to begin to probe, the significance of labor market conditions for the process of suburbanization.

Although national and local conditions mattered, the greatest influence on whether workers would follow an employer into the suburbs was the security of the employment situation they would find there. No company could offer jobs for life. In the face of uncertainty, the safest bet was for workers to settle in or near an industrial suburb that contained a number of potential employers. In the case of skilled workers, these might have to be in a particular industry. Indeed, one of the reasons why such districts were developed is that they were effective in attracting and retaining skilled workers. Particularly if they were specialized in just one or two industrial sectors, even the largest industrial suburbs offered less job security than the city itself. The situation of workers in more isolated settings could be vulnerable indeed, as Gunter Gad shows in the case of several of Toronto's suburban factories during the 1930s and 1940s. The safest bets were large plants operated by major companies, such as Caterpillar from the 1930s onward, but the experience of Pullman dramatized the fact that even these were vulnerable to economic downturns. The prospects of small companies being able to draw workers to isolated suburban sites were small indeed, and that is surely the reason why few small companies attempted such a strategy.

A fifth, and perhaps the most obvious, influence on the suburban dynamic is the availability of effective transportation, including mass transit and the private automobile. It is a deceptive truism that the growth of residential suburbs depends on the availability of affordable, and fairly rapid, transportation. The stereotypical examples are commuter railroad suburbs such as Riverside, streetcar suburbs such as those described by Sam Bass Warner, and the automobile suburbs that grew up around every city after the 1920s.[35] Cities like New York, and to a lesser extent Chicago, that were favored with commuter railroads and mass transit saw a disproportionate amount of residential suburbanization. Where transit development was more constrained, industrial or balanced suburbanization played a larger role. This logic makes most sense for the very largest metropolitan areas, such as Pittsburgh and Los Angeles. It does not necessarily illuminate the experience of smaller places. In Toronto, for example, until 1921 the company that held the streetcar franchise refused to extend lines into the suburbs. This did not prevent the subdivision and settlement of residential suburbs, however, since commuters were willing to walk up to a mile, or even more, to the end of the streetcar line. Commuter railroads and streetcars, then, were neither a necessary nor a sufficient condition for residential suburbanization. They simply made it more likely.

Trends

The dynamic of the work-home linkage evolved, and in a systematic fashion. As Richard Walker and Robert Lewis note in Chapter 2, it was once conventional to assume that industrial decentralization began around the turn of the century and gathered pace during the 1940s. This view implied that residential suburbanization was originally the dominant force. The papers collected here show that the history of industrial decentralization is much longer than has commonly been supposed. But it is true that the middle decades of the twentieth century saw some significant changes in the scale of suburbanization. It is possible that it was residential suburbanization that became more common in this period. At any rate, there can be no doubt that the importance of home-work linkages became less clear as they were more attenuated. Two developments helped to encourage these trends.

The first long-term development to shape suburbanization has been the growing importance of women, especially married women, in the labor force. This trend eventually helped to make suburbs more attractive to employers, although it is not clear when. As late as 1960, a study in Chicago found that women were reluctant to take suburban jobs, and for a time this may have discouraged potential employers from relocating.[36] The growing participation of women in the labor force increased the size of the suburban labor pool and eventually helped to draw into the suburbs many companies that employ women. In recent decades this has been discussed as a factor in the location of offices. Many women have always been engaged in factory work, however, and it is likely that the location of manufacturing has also been influenced by the secular growth of women's employment.

A second and more significant shift for the suburbanization process has been a steady increase in the distance between home and workplace, and flexibility in the patterning of the employment linkage. The basic facts are widely appreciated, but all of their implications are not.[37] When people walked to work, the suburbanization of homes and workplaces was tightly linked. Residential districts had to be close to the centers of employment, whether in the city or adjacent suburbs. Both employers and workers made careful calculations about the risks of moving to more remote locations. Fixed-rail transit made it possible for people to travel further to work. It reinforced the accessibility advantage of the central city, albeit by encouraging residential suburbs to grow up around train stations or along streetcar lines.[38]

Subsequently, the automobile loosened everything up. Cars made longer-distance commuting possible for a majority of workers, which, as in Flint, encouraged the growth of residential suburbs. More important, cars revolutionized the relationship between suburbs. A ten-mile drive from one suburb to another took less time than the same commute into the city, and parking was easier. Because the automobile altered the relationships between suburbs, it changed what a suburb could be. In the early twentieth century it was both easy and meaningful to distinguish between a residential suburb such as Riverside, Illinois, and a nearby balanced-industrial suburb such as Melrose Park. It would have been very difficult to commute from one to the other since they were poorly linked by transit and too far apart to walk between. In some cases the self-containment of industrial and balanced suburbs persisted into the early postwar period. In Framingham, Massachusetts, a town of only 25,000 in 1945, workers' homes were still clustered around each of three large mills.[39] In East Chicago, a larger community on the fringe of a very much larger metropolitan area, in the late 1940s about half of all of the blue-collar workers employed at the main steel plant lived within walking distance of work.[40] Factories that were established at the suburban fringe could draw on an extensive region, but with the prospect of a job many families were still quite keen to move close to work. In the Los Angeles area, soon after Kaiser Steel moved to Fontana in 1947, barely 30 percent of its workers lived in the community; a year later more than 40 percent of a larger workforce lived locally.[41] At Willow Run a similar "tightening" of the labor shed had taken place even within the first six months of operation.[42] People did not always use cars and commute long distances just because they could afford to do so.

By the 1950s, however, the widespread availability of the car meant that older suburbs such as Melrose Park and Riverside, along with newer postwar suburbs that were also just a few miles apart, could easily be linked.

Of course, many of the postwar suburbs developed on a larger scale than their predecessors. The 1950s saw the development of industrial, and of wholly residential, districts that dwarfed anything that had gone before. Hise emphasizes this change in scale in Los Angeles, for example, when he contrasts the Leimert Park development of the 1920s, which saw the creation of 1,200 lots on 230 acres, with the postwar project of Panorama City, which included 3,000 dwelling units on 800 acres, with a regional shopping center thrown in for good measure. Even more extreme examples were the three Levittowns, which were not only the largest projects to be planned and developed by a single entrepreneur but also probably the largest to be exclusively residential in character.[43] Clearly, the automobile encouraged developers to design projects that segregated land uses on an unprecedented scale.

Although the scale of land-use segregation increased, it did not keep pace with the potential of the new mode of transportation. Vance gives us some idea of the expanding possibilities. Using city directories he has documented the growth of the employment field of Natick, Massachusetts, that is to say, the area within which those who lived in the town were able and willing to commute. As a walking city in the mid-nineteenth century, Natick boasted an employment field of about fifteen square miles.[44] By the turn of the twentieth century the street railway had expanded this tenfold. By 1950 80 percent of Natick's employed residents commuted by car, and the employment field had increased to almost 1,200 square miles, an eighty-fold increase from the walking era. In the postwar era commuters could disperse across a far-flung employment field; industrial districts could draw upon a widening suburban and exurban labor shed. Some suburbs might contain a roughly equal number of workers and jobs, but it was no longer even conceivable that a majority of the local jobs would be filled by local workers. Don Mills is a case in point. This planned suburb north of Toronto was developed by a single entrepreneur in the early 1950s. It was designed to be self-contained and in theory could have been: in addition to single-family homes for middle-income families it included a substantial number of rental units in low-rise apartments, a shopping center, and extensive office and industrial districts that attracted businesses such as IBM.[45] But, from the beginning, most residents commuted elsewhere and most of those who worked in the area arrived daily from beyond its boundaries. What, then, are we to make of places like Panorama City, which, as Hise describes it, was marketed in the 1940s on the basis of its location next to several major employers, including General Motors and Jergens? How important was this proximity for the first residents? Did the

area ever really function as part of a balanced unit of suburban development, or was the very concept of balanced suburbanization ceasing to have
meaning? In truth, although people referred to Don Mills and Panorama
City as suburbs, these places had little or no functional identity.

The rise of the automobile created a new scale and flexibility of the
home-work linkage that undermined the tidy logic by which suburbs had
been labeled. Panorama City was itself mostly residential in character, but
functionally it was part of a larger regional pattern of balanced suburban
development. The same trend was happening everywhere, and the postwar era saw the disappearance of the truly residential suburb.[46] By extension, the same argument applies to all types of suburbs: the distinction
between industrial, residential, and balanced suburbs ceased to have meaning. We should not at this point simply abandon all hope of discerning
causality. It is still meaningful to distinguish between the processes of
industrial and residential suburbanization. New factories in far-flung locations still drew people ever outward. The establishment of Ford's assembly plant at Oakville, for example, more than twenty miles from downtown
Toronto, gave perceptible impetus to the westward growth of the metropolitan area in the 1960s and 1970s. But Ford's capacity to tap labor from
Burlington and even Hamilton to the west, as well as from Mississauga and
Toronto to the east, underlines the growing difficulty of distinguishing
cause and effect.

Discussion

The argument that I have developed in this chapter could have been
framed as a series of hypotheses about why the suburbanization process
has taken different forms in different urban areas. Indeed, perhaps the
comments about long-run trends should have been framed this way in
order to emphasize their speculative character. The truth is that we know
very little about how the employment linkage has varied and changed, and
what role it has played in the suburban trend.

If we are going to develop a clearer idea of the importance of the home-
work linkage we will need to probe beneath the aggregate information that
is most readily available. More often than we might assume, it is possible
to construct a plausible account from published sources. A good example
may be found in Becky Nicolaides's recent study of South Gate, an early
twentieth-century suburb adjacent to what is now known as South Central Los Angeles.[47] Nicolaides uses census and related sources to show
that, from its inception, the area contained a substantial number of both

homes and workplaces. In 1940, for example, it contained an almost exactly equal number of each. Other data, however, show that it would be misleading to assume that South Gate was the product of balanced suburbanization. Transit surveys indicate that even in the 1920s it was not self-contained, either as a labor shed or as an employment field. The best available data are for 1940, when survey data compiled by the Federal Writers Project indicated that at most 29 percent of the people employed in South Gate factories actually lived in the area, while about the same proportion of blue-collar residents of the area actually worked locally.[48] Unfortunately, none of these data show how many of South Gate's residents commuted in the direction of downtown, as opposed to other suburbs. It may be that this district never functioned primarily as a suburb at all. Certainly its development was always tied very closely to that of adjacent industrial and residential districts.

To supplement existing surveys, and where none exist, we may sometimes be able to reconstruct home-work linkages from city directories.[49] Until at least the 1940s, directories routinely included information on the occupation of most employed adults, and sometimes their employers' names too. The task of combing the directories for the names and addresses of those who worked for particular employers can be daunting and is fraught with difficulties that I have discussed elsewhere.[50] The potential of this source, however, is enormous. Available annually, they can support sensitive analyses of short-run change. For example, they make it possible to determine whether the first residents of a residential suburb were employed downtown or in a neighboring suburb. As I indicated above with reference to Kodak, Canada, they also enable us to calculate what proportion of the employees of a decentralizing factory remain with the company and, perhaps, follow it out to the suburban fringe. In the right circumstances they make it possible to analyze the process of suburbanization with rewarding precision.

It takes a lot of work to glean useful information from city directories. Readers who have followed the argument this far may wonder whether the effort would really be worthwhile. As someone who has made that effort I have a stake in overstating the case, so let me be clear. There are situations where the broad outline of the suburbanization process can be inferred from the patterns that it produced on the ground. It would surely be overkill to comb directories in order to show that early residents of Riverside commuted to Chicago, or that the first mill workers living in Homestead were employed locally. Then again, it is often possible to make reasonable inferences from fragmentary sources, including surveys and newspaper accounts.

For example, a telling piece of information about Leaside in its early years, when it was a small industrial suburb of Toronto, is contained in a contemporary newspaper report that its first large employer had to run special trains to bring workers from downtown. Obviously it makes sense to seek out and exploit to the full these sorts of information.

In many cases, however, it is simply not possible to make precise and reasonable inferences about the home-work linkages without investing substantial effort in gathering data. What is the payoff? There is a certain intellectual satisfaction to be derived from being able to show, as opposed to infer or guess, how a place developed. It is an intellectual satisfaction that gains resonance when local experiences are seen in the sort of wider comparative and historical frame of reference that I have outlined here. There are also other, perhaps more important, issues at stake. If we knew more about exactly how the decentralization of homes and industry played off each other, we might discover something about the calculations and priorities of the agents involved. If we knew what proportion of their labor force companies usually lost when they relocated to the fringe, for example, and how frequently they made that kind of relocation decision, that should tell us something about the way those companies viewed their labor force. If we discovered that a small company moving into an industrial suburb soon began to tap local labor, perhaps drawing workers away from other local employers, we would have confirmation that the area functioned as an industrial district. If we knew whether a residential suburb housed downtown commuters, as opposed to workers in adjacent suburbs, we might learn about the tradeoffs families were willing to make, about how they viewed suburban living in general, and about what sorts of commitments they felt for their place of residence. These are some of the judgments that Becky Nicolaides tries to make in her study of South Gate, suggesting that weak ties between homes and local workplaces shaped class and municipal politics.[51]

The employment linkage connects home and work, and historical research on the subject offers the prospect of connecting the issues that have commonly been addressed by historians of business, labor, and the family. Such research would also bring into a more fruitful relationship the established literature on residential suburbs with the growing body of work on suburban industry. It promises, then, to help us to see cities and city lives as they are experienced, as complicated wholes.

Notes

Chapter One

1. M. Adams, "Present Housing Conditions in South Chicago, South Deering and Pullman" (master's thesis, University of Chicago, 1926); S. Buder, *Pullman: An Experiment in Industrial Order and Community Planning, 1880–1930* (New York, 1967); D. Pacyga, *Polish Immigrants and Industrial Chicago: Workers on the South Side, 1880–1922* (Columbus, 1991).

2. D. Lindstrom, *Economic Development in the Philadelphia Region, 1810–1850* (New York, 1978); P. Scranton, *Proprietary Capitalism: The Textile Manufacture at Philadelphia, 1800–1885* (New York, 1983); D. Bierne, "Hampden-Woodberry: The Mill Village in an Urban Setting," in *Baltimore: A Perspective on Historical Urban Development*, ed. S. Bennett and C. Christian (Baltimore, 1989).

3. G. Taylor, *Satellite Cities: A Study of Industrial Suburbs* (New York, 1915); Buder, *Pullman*.

4. For a list of suburbs with an industrial base in 1940, see G. Kneedler, "Economic Classification of Cities," in *The Municipal Year Book* (Chicago, 1945). Also see R. Harper, "Metro East: Heavy Industry in the St. Louis Metropolitan Area (no publication information); McClellan and Junkersfeld, Inc., *Report on Transportation in the Milwaukee Metropolitan District to the Transportation Survey Committee of Milwaukee*, vol. 1 (New York, 1928); A. Orum, *City Building in America* (Boulder, 1995); L. Thomas, *The Localization of Business Activities in Metropolitan St. Louis* (St. Louis, 1927); M. Goldman, *High Hopes: The Rise and Decline of Buffalo, New York* (Albany, 1983); O. Zunz, *The Changing Face of Inequality: Urbanization, Industrial Development, and Immigrants in Detroit, 1880–1920* (Chicago, 1982).

5. B. Berry and Y. Cohen, "Decentralization of Commerce and Industry," in *The Urbanization of the Suburbs*, ed. L. Masotti and J. Haddon (Beverly Hills, 1973), 431–55; P. Muller, *Contemporary Suburban America* (Englewood Cliffs, 1981); L. Schnore, "Metropolitan Growth and Decentralization," *American Journal of Sociology* 63 (1957): 171–80; A. Scott, "Production System Dynamics and Metropolitan Development," *Annals of the Association of American Geographers* 72 (1982): 185–200; D. Slater, "Decentralization of Urban Peoples and Manufacturing Activity in Canada," *Canadian Journal of Economics and Political Science* 27 (1961): 72–84.

6. S. B. Warner, *Streetcar Suburbs: The Process of Growth in Boston, 1870–1900* (Cambridge, 1962). A sample of the more recent work includes John Weaver, "From Land Assembly to Social Maturity: The Suburban Life of Westdale (Hamilton), Ontario, 1911–1951," *Historie Sociale/Social History* 11 (1978): 411–40; C. O'Connor, *A Sort of Utopia: Scarsdale, 1891–1981* (Albany, 1983); K. Jackson, *Crabgrass Frontier: The Suburbanization of the United States* (New York, 1985); R. Fishman, *Bourgeois Utopias: The Rise and Fall of Suburbia* (New York, 1987); M. Ebner, *Creating Chicago's North Shore* (Chicago, 1988); M. Weiss, *The Rise of the Community Builders: The American Real Estate Industry and Urban Planning* (New York, 1987); J. Stilgoe, *Borderland: Origins of the American Suburb, 1820–1939* (New Haven, 1988); A. D. Keating, *Building Chicago: Suburban Developers and the Creation of a Divided Metropolis* (Columbus, 1988); M. Marsh, *Suburban Lives* (New Brunswick, 1990); A. Von Hoffman, *Local Attachments: The Making of an American Urban Neighborhood, 1850 to 1920* (Baltimore, 1994); L. McCann, "Planning and Building the Corporate Suburb of Mont Royal, 1910–1925," *Planning Perspectives* 11 (1996): 259–301; G. Hise, *Magnetic Los Angeles: Planning the Twentieth-Century Metropolis* (Baltimore, 1997); M. Sies, "North American Suburbs, 1880–1950: Cultural and Social Reconsiderations," *Journal of Urban History* 27 (2001): 313–46; P. Mattingly, *Suburban Landscapes: Culture and Politics in a New York Metropolitan Community* (Baltimore, 2001).

7. G. McLaughlin, *Growth of American Manufacturing Areas: A Comparative Analysis with Special Emphasis on Trends in the Pittsburgh District* (Pittsburgh, 1938); D. Creamer, *Is Industry Decentralizing? A Statistical Analysis of Locational Changes in Manufacturing Employment, 1899–1933* (Philadelphia, 1935). See also W. Mitchell, *Trends in Industrial Location in the Chicago Region since 1920* (Chicago, 1933); G. Martin, "Étude des factuers qui ont déterminé la localisation de l'industrie à Montréal et des banlieues," *Revue Trimestrielle Canadienne* 34 (1934): 297–335.

8. C. Woodbury and F. Cliffe, "Industrial Location and Urban Redevelopment," in *The Future of Cities and Urban Redevelopment*, ed. C. Woodbury (Chicago, 1953), 103–288. See also E. Kitagawa and D. Bogue, *Suburbanization of Manufacturing Activity within Standard Metropolitan Areas* (Oxford, Ohio, 1955); P. Reid, *Industrial Decentralization: Detroit Region, 1940–1950* (Detroit, 1951); Schnore, "Metropolitan Growth and Decentralization"; W. Zelinksy, "Has America Been Decentralizing? The Evidence for the 1939–1954 Period," *Economic Geography* 38 (1962): 251–69; Slater, "Decentralization of Urban Peoples."

9. Taylor, *Satellite Cities*; M. Byington, *Homestead: The Households of Mill Town* (New York, 1910; 1969); Regional Plan of New York and Its Environs, *Regional Survey of New York and Its Environs*, vols. 1a and 1b (New York, 1928; 1974); E. E. Pratt, *Industrial Causes of Congestion of Population in New York City* (New York, 1911; 1968). See also E. Muller, "The Pittsburgh Survey and 'Greater Pittsburgh': A Muddled Metropolitan Geography," in *Pittsburgh Surveyed: Social Science and Social Reform in the Early Twentieth Century*, ed. M. Greenwald and M. Anderson (Pittsburgh, 1996), 69–87.

10. H. Douglass, *The Suburban Trend* (New York, 1925).

11. E. Muller and P. Groves, "The Emergence of Industrial Districts in Mid-Nineteenth-Century Baltimore," *Geographical Review* 69 (1979): 159–78 (republished with minor revisions as Chapter 3 in this volume); Buder, *Pullman*; J. Kenyon, *Industrial Localization and Metropolitan Growth: The Paterson-Passaic District* (Chicago, 1960); J. Kenyon, *The Industrialization of the Skokie Area* (Chicago, 1954); Zunz, *Changing Face*

of Inequality; R. Cramer, *Manufacturing Structure of the Cicero District, Metropolitan Chicago* (Chicago, 1952); E. Hoover and R. Vernon, *Anatomy of a Metropolis: The Changing Distribution of People and Jobs within the New York Metropolitan Region* (New York, 1959). One of the few earlier studies to address the impact of industrial decentralization on suburban residence and industry is H. Binford, *The First Suburbs: Residential Communities on the Boston Periphery, 1815–1860* (Chicago, 1985).

12. For an example of sociological work on the working-class suburb, see B. Berger, *Blue Collar Suburb: A Study of Auto Workers in Suburbia* (Berkeley and Los Angeles, 1968).

13. D. Gordon, "Capitalist Development and the History of American Cities," in *Marxism and the Metropolis*, ed. W. Tabb and L. Sawers (New York, 1984), 113–38; E. Greer, "Monopoly and Competitive Capital in the Making of Gary, Indiana," *Science and Society* 40 (1976): 465–78; P. O'Donnell, "Industrial Capitalism and the Rise of Modern American Cities," *Kapitalistate* 5 (1977): 91–128; L. Sawers, "Urban Form and the Mode of Production," *Review of Radical Political Economies* 7 (1975): 52–68; R. Walker, "The Transformation of Urban Structure in the Nineteenth Century and the Beginnings of Suburbanization," in *Urbanization and Conflict in Market Societies*, ed. K. Cox (Chicago, 1978), 165–212; R. Walker, "A Theory of Suburbanization: Capitalism and the Construction of Urban Space in the United States," in *Urbanization and Urban Planning in Capitalist Society*, ed. M. Dear and A. Scott (New York, 1981), 383–429.

14. A. Scott, "Locational Patterns and Dynamics of Industrial Activity in the Modern Metropolis," *Urban Studies* 19 (1982): 111–42; A. Scott, "Industrial Organization and the Logic of Intra-Metropolitan Location: I. Theoretical Considerations," *Economic Geography* 59 (1983): 233–50; A. Scott, *Metropolis: From the Division of Labor to Urban Form* (Los Angeles, 1988); M. Storper and A. Scott, "The Geographical Foundations and Social Regulation of Flexible Production Complexes," in *Territory and Social Reproduction*, ed. J. Wolch and M. Dear (Boston, 1988), 21–40.

15. R. Harris, *Unplanned Suburbs: Toronto's American Tragedy, 1900 to 1950* (Baltimore, 1996); R. Harris and M. Sendbuehler, "The Making of a Working-Class Suburb in Hamilton's East End, 1900–1945," *Journal of Urban History* 20 (1994): 486–511; H. Silcox, *A Place to Live and Work: The Henry Disston Saw Works and the Tacony Community of Philadelphia* (University Park, Pa., 1994); K. Kane and T. Bell, "Suburbs for a Labor Elite," *Geographical Review* 75 (1985): 319–34; H. McKiven, *Iron and Steel: Class, Race, and Community in Birmingham, Alabama, 1875–1920* (Chapel Hill, 1995); B. Nicolaides, *My Blue Heaven: Life and Politics in the Working-Class Suburbs of Los Angeles, 1920–1965* (Chicago, 2002); A. Wiese, "The Other Suburbanites: African American Suburbanization in the North before 1950," *Journal of American History* 85 (1999): 1–26; P.-A. Linteau, *The Promoter's City: Building the Industrial Town of Maisonneuve, 1883–1918* (Toronto, 1985); T. Gardner, "The Slow Wave: The Changing Residential Status of Cities and Suburbs in the United States, 1850–1940," *Journal of Urban History* 27 (2001): 293–312.

16. Excluding the work in this volume, some of this research includes R. Harris, "Industry and Residence: The Decentralization of New York City, 1900–1940," *Journal of Historical Geography* 19 (1993): 169–90; Linteau, *Promoter's City*; A. Mosher, "'Something Better Than the Best': Industrial Restructuring, George McMurtry, and the Creation of the Model Industrial Town of Vandergrift, Pennsylvania, 1883–1901," *Annals of the Association of American Geographers* 85 (1995): 84–107; F. Viehe, "Black

Gold Suburbs: The Influence of the Extractive Industry on the Suburbanization of Los Angeles, 1890–1930," *Journal of Urban History* 8 (1981): 3–26.

17. For recent overviews, see R. Lewis, "Running Rings around the City: North American Industrial Suburbs, 1850–1950," in *Changing Suburbs: Foundation, Form, and Function*, ed. R. Harris and P. Larkham (London, 1999), 146–67; R. Harris and R. Lewis, "The Geography of North American Cities and Suburbs, 1900–1950: A New Synthesis," *Journal of Urban History* 27 (2001): 262–92; J. Wunsch, "The Suburban Cliché," *Social History* (1995): 643–58.

18. The suburban employment data was collected by the author from various censuses and is summarized in R. Lewis, "The Changing Fortunes of American Central-City Manufacturing, 1870–1950," *Journal of Urban History* 28 (2002): 573–98.

19. Other notable examples of writers emphasizing the cyclical nature of city building are S. Olson, *Baltimore: The Building of an American City* (Baltimore, 1980); W. Isard, "A Neglected Cycle: The Transport-Building Cycle," *Review of Economic and Statistics* 24 (1942): 149–58; H. Hoyt, *One Hundred Years of Land Values in Chicago* (Chicago, 1933); J. Whitehead, *The Changing Face of Cities: A Study of Development Cycles and Urban Form* (New York, 1987).

20. R. Blaszczyk, "'Reign of the Robots': The Homer Laughlin China Company and Flexible Mass Production," *Technology and Culture* 36 (1995): 863–911; D. Hiebert, "Discontinuity and the Emergence of Flexible Production: Garment Production in Toronto, 1901–1931," *Economic Geography* 66 (1990): 229–53; P. Hirst and J. Zeitlin, "Flexible Specialization versus Post-Fordism: Theory, Evidence, and Policy Implications," *Economy and Society* 20 (1991): 1–56; B. Laurie and M. Schmitz, "Manufacture and Productivity: The Making of an Industrial Base, Philadelphia, 1850–1880," in *Philadelphia: Work, Space, Family, and Group Experience in the Nineteenth Century*, ed. T. Hershberg (New York, 1981), 43–92; R. Lewis, "Production and Spatial Strategies in the Montreal Tobacco Industry, 1850–1918," *Economic Geography* 70 (1994): 370–89; W. Licht, *Getting Work: Philadelphia, 1840–1950* (Cambridge, 1992); P. Scranton, *Proprietary Capitalism: The Textile Manufacture at Philadelphia, 1800–1885* (New York, 1983); P. Scranton, *Endless Novelty: Specialty Production and American Industrialization, 1865–1925* (Princeton, 1997); P. Scranton and W. Licht, *Work Sights: Industrial Philadelphia, 1890–1950* (Philadelphia, 1986).

21. For other studies that discuss the rural-urban relationship, see G. Brechin, *Imperial San Francisco: Urban Power, Earthly Ruin* (Berkeley, 1999); W. Cronon, *Nature's Metropolis: Chicago and the Great West* (New York, 1991); W. Robbins, *Colony and Empire: The Capitalist Transformation of the American West* (Lawrence, Kans., 1994).

22. Harris, *Unplanned Suburbs*; Nicolaides, *My Blue Heaven*.

23. Lewis, "Changing Fortunes"; J. Teaford, *City and Suburb: The Political Fragmentation of Metropolitan America, 1850–1970* (Baltimore, 1979); Jackson, *Crabgrass Frontier*, chap. 8.

24. E. W. Burgess, "The Growth of the City," in *The City*, ed. R. Park, E. W. Burgess, and R. D. McKenzie (Chicago, 1925), 47–62; L. Schnore, "The Socio-economic Status of Cities and Suburbs," *American Sociological Review* 28 (1963): 76–85; Warner, *Streetcar Suburbs*. Although not always explicitly stated, most writers have assumed, often against their better judgment, that this gradient was in place. See R. Harris and R. Lewis, "Constructing a Fault(y) Zone: Misrepresentations of American Cities and Suburbs, 1900–1950," *Annals of the Association of American Geographers* 88 (1998): 622–39.

25. J. Reiff, "'His statements . . . will be challenged': Ethnicity, Gender, and Class in the Evolution of the Pullman-Roseland Area of Chicago, 1894–1917," *Mid-America* 74 (1992): 231–52; J. Reiff, "Rethinking Pullman: Urban Space and Working-Class Activism," *Social Science History* 24 (2000): 7–32; Bierne, "Hampden-Woodberry."

26. R. Lewis, *Manufacturing Montreal: The Making of an Industrial Landscape, 1850 to 1930* (Baltimore, 2000), 237–44; Normand Mousette, *En ces lieux que l'on nomma 'La Chine'* (Lachine, 1978).

27. Kenyon, *Industrial Localization and Metropolitan Growth.*

Chapter Two

Acknowledgments: Dedicated to the memory of David Gordon (1945–96), whose insights on industrial decentralization, penned years ago, are part of the genesis of the present work. Many thanks for the input of David Meyer, Greg Hise, Ted Muller, and Mary Beth Pudup.

Epigraph: Quoted in C. Clark, "Land Subdivision," in *Los Angeles: Preface to a Master Plan,* ed. G. Robbins and D. Tilton (Los Angeles, 1941).

1. S. B. Warner, *Streetcar Suburbs: The Process of Growth in Boston, 1870–1900* (Cambridge, 1962); K. Jackson, *Crabgrass Frontier: The Suburbanization of the United States* (New York, 1985); R. Fishman, *Bourgeois Utopias: The Rise and Fall of Suburbia* (New York, 1987); J. Stilgoe, *Borderland: Origins of the American Suburb, 1820–1939* (New Haven, 1988); R. Walker, "The Transformation of Urban Structure in the Nineteenth Century and the Beginnings of Suburbanization," in *Urbanization and Conflict in Market Societies,* ed. K. Cox (Chicago, 1978), 165–213.

2. We realize that many upper- and middle-class residential suburbs were built away from industry and not in concert with it and that any complete theory of suburbanization must include both movements.

3. On the debates on congestion, urban ills, and the benefits of single-family suburbia, see R. Walker, "The Suburban Solution: Capitalism and the Construction of Urban Space in the United States" (Ph.D. diss., Johns Hopkins University, 1977); G. Wright, *Building the Dream: a Social History of Housing in America* (New York, 1981); P. Groth, *Living Downtown: The History of Residential Hotels in the United States* (Berkeley, 1994); E. Muller, "The Pittsburgh Survey and 'Greater Pittsburgh': A Muddled Metropolitan Geography," in *Pittsburgh Surveyed: Social Science and Social Reform in the Early Twentieth Century,* ed. M. Greenwald and M. Anderson (Pittsburgh, 1996), 69–87; P. Rutherford, *Saving the Canadian City: The First Phase, 1880–1920* (Toronto, 1974); J. Weaver, *Shaping the Canadian City: Essays on Urban Politics and Policy, 1890–1920* (Toronto, 1977). For some contemporary views, see H. Ames, *The City Below the Hill* (Toronto, 1897; 1972); E. E. Pratt, *Industrial Causes of Congestion of Population in New York City* (New York, 1911); G. Taylor, *Satellite Cities: A Study of Industrial Suburbs* (New York, 1915); P. Kellogg, ed., *The Pittsburgh Survey,* 6 vols. (New York, 1910–16).

4. R. Park, E. W. Burgess, and R. D. McKenzie, eds., *The City* (Chicago, 1925); R. D. McKenzie, *The Metropolitan Community* (New York, 1931). For a Canadian example, see C. Dawson, "The City as an Organism," *La Revue Municipale* (1927): 11–12. For critiques of the Park-Burgess school of urban ecology, see D. Harvey, *Social Justice and the City* (Baltimore, 1973); M. Castells, *The Urban Question* (London, 1975).

5. H. Douglass, *The Suburban Trend* (New York, 1925).

6. R. Haig, ed., *Regional Survey of New York and Its Environs*, vol. 1, *Major Economic Factors in Metropolitan Growth and Arrangement* (New York, 1927). For Canada, see G. Ferguson, "Decentralization of Industry and Metropolitan Control," *Journal of the Planning Institute* 2 (1923): 5–12. As far as we are aware, other than Ferguson, there are no comparable studies of Canadian cities before World War II. On the key role of the RPA in New York planning, see R. Caro, *The Power Broker: Robert Moses and the Fall of New York* (New York, 1974); M. Heiman, *The Quiet Evolution: Power, Planning, and Profits in New York State* (New York, 1988).

7. H. Hoyt, *One Hundred Years of Land Values in Chicago* (Chicago, 1933), and *The Structure and Growth of Residential Neighborhoods in American Cities* (Washington, D.C., 1937); National Resources Committee, *Our Cities: Their Role in the National Economy* (Washington, D.C., 1937); D. Creamer, *Is Industry Decentralizing? A Statistical Analysis of Locational Changes in Manufacturing Employment, 1899–1933* (Philadelphia, 1935); G. McLaughlin, *Growth of American Manufacturing Areas: A Comparative Analysis with Special Emphasis on Trends in the Pittsburgh District* (Pittsburgh, 1938); J. Delage, "L'industrie manufacturière," in *Montréal Economique*, ed. E. Minville (Montreal, 1943), 183–241.

8. On the war years, see J. Kain, "The Distribution and Movement of Jobs and Industry," in *The Metropolitan Enigma*, ed. J. Wilson (Cambridge, 1968), 1–43; see also C. Woodbury, ed., *The Future of Cities and Urban Redevelopment* (Chicago, 1953).

9. Editors of *Fortune*, *The Exploding Metropolis* (New York, 1958); W. Whyte, *The Organization Man* (New York, 1956). This literature continued through the property boom up to the crisis of 1975. W. Whyte, *The Last Landscape* (Garden City, 1968); M. Clawson, *Suburban Land Conversion in the United States: An Economic and Governmental Process* (Baltimore, 1968); R. Fellmeth, ed., *The Politics of Land* (New York, 1973). On Canada, see R. Blanchard, *L'ouest du Canada française: Montréal et sa Region* (Montreal, 1953); S. Clark, *The Suburban Society* (Toronto, 1966).

10. R. Beauregard, *Voices of Decline: The Postwar Fate of US Cities* (Cambridge, 1995).

11. E. Hoover and R. Vernon, *Anatomy of a Metropolis* (Cambridge, 1959); R. Vernon, *Metropolis 1985: An Interpretation of the Findings of the New York Metropolitan Region Study* (Cambridge, 1960); and the accompanying studies, B. Chinitz, *Freight in the Metropolis* (Cambridge, 1960); M. Segal, *Wages in the Metropolis: Their Influence on the Location of Industries in the New York Region* (Cambridge, 1960). Hoover had already stated the theory of industrial dispersal as a result of industrial maturation in *The Location of Economic Activity* (New York, 1948).

12. For example, D. Slater, "Decentralization of Urban Peoples and Manufacturing Activity in Canada," *Canadian Journal of Economics and Political Science* 27 (1961): 72–84; W. Alonso, *Location and Land Use* (Cambridge, 1964); A. Pred, "The Intrametropolitan Location of American Manufacturing," *Annals of the Association of American Geographers* 54 (1964): 165–80; L. Moses and H. Williamson, "The Location of Economic Activity in Cities," *American Economic Review, Papers and Proceedings* (1967): 211–22; E. Mills, *Studies in the Structure of the Urban Economy* (Baltimore, 1972); A. Hamer, *Industrial Exodus from the Central City* (Lexington, Mass., 1973).

13. D. Gordon, "Capitalist Development and the History of American Cities," in *Marxism and the Metropolis*, ed. W. Tabb and L. Sawers (New York, 1984), 21–53; J. Vance, *This Scene of Man* (New York, 1977); J. Teaford, *The Unheralded Triumph: City Government in America, 1870–1900* (Baltimore, 1984); E. Muller, "From Waterfront

to Metropolitan Region: The Geographical Development of American Cities," in *American Urbanism: a Historiographic Review*, ed. H. Gillette and Z. Miller (Westport, 1987), 105–133; R. Lewis, "Running Rings around the City: North American Industrial Suburbs, 1850–1950," in *Changing Suburbs: Foundation, Form, and Function*, ed. R. Harris and P. Larkham (London, 1999), 146–67.

14. Harvey, *Social Justice and the City*; Castells, *Urban Question*; P. Muller, *Contemporary Suburban America* (Englewood Cliffs, 1981); Walker, "Suburban Solution," "Transformation of Urban Structure," and "Theory of Suburbanization"; A. Scott, "Locational Patterns and Dynamics of Industrial Activity in the Modern Metropolis," *Urban Studies* 19 (1982): 111–142, "Industrial Organization and the Logic of Intrametropolitan Location: I. Theoretical Considerations," *Economic Geography* 59 (1983): 233–50, and "Industrialization and Urbanization: A Geographic Agenda," *Annals of the Association of American Geographers* 76 (1986): 25–37; Jackson, *Crabgrass Frontier*; Stilgoe, *Borderland*; Fishman, *Bourgeois Utopias*. For Canada, see G. Stelter, "The City-Building Process in Canada," in *Shaping the Urban Landscape*, ed. G. Stelter and A. Artibise (Ottawa, 1982), 1–29.

15. Fishman, *Bourgeois Utopias*; Muller, *Contemporary Suburban America*; E. Soja, *Post-modern Geographies* (London, 1989); J. Garreau, *Edge City: Life on the New Frontier* (New York, 1991); R. Kling, S. Olin, and M. Poster, eds., *Postsuburban California: The Transformation of Orange County since World War II* (Los Angeles, 1991); T. Stanback, *The New Suburbanization: Challenge to the Central City* (Boulder, 1991).

16. Taylor, *Satellite Cities*; Creamer, *Is Industry Decentralizing?*; McLaughlin, *Growth of American Manufacturing*; Woodbury, *Future of Cities*; Kain, "Distribution and Movement of Jobs."

17. A. Scott, "Production System Dynamics and Metropolitan Development," *Annals of the Association of American Geographers* 72 (1982): 188.

18. A. Weber, *Theory of the Location of Industries* (Chicago, 1908; 1929). Also see Hoover, *Location of Economic Activity*.

19. Moses and Williamson, "Location of Economic Activity"; Mills, *Structure of the Urban Economy*.

20. Jackson, *Crabgrass Frontier*, 184. Chinitz, *Freight in the Metropolis*, and B. Chinitz and R. Vernon, "Changing Forces in Industrial Location," *Harvard Business Review* 38 (1962): 126–36; Hamer, *Industrial Exodus*; Slater, "Decentralization of Urban Peoples."

21. Hoover and Vernon, *Anatomy of a Metropolis*; Pred, "Intrametropolitan Location"; Hamer, *Industrial Exodus*; P. Leone and R. Struyk, "The Incubator Hypothesis: Evidence from Five SMSAs," *Urban Studies* 13 (1976): 325–31. The theory of industrial maturation was first put forward by S. Kuznets, *Secular Movements in Production and Prices* (Boston, 1930), and A. Burns, *Measuring Business Cycles* (New York, 1932).

22. Scott, "Locational Patterns" and "Production System Dynamics."

23. Ibid. H. Watts, *The Large Industrial Enterprise* (London, 1980), and *The Branch Plant Economy* (London, 1981). Highly influential was the model of international corporate hierarchy of S. Hymer, "The Multinational Corporation and the Law of Uneven Development," in *Economics and World Order*, ed. J. Bhagwati (New York, 1972), 113–40.

24. Taylor, *Satellite Cities*; Hoyt, *Structure and Growth*.

25. A. Fishlow, *American Railroads and the Transformation of the Ante-Bellum Economy* (Cambridge, 1962); H. Scheiber, *Ohio Canal Era* (Athens, 1969); M. Pudup, "Packers and Reapers, Merchants and Manufacturers: Industrial Structuring and Location

in an Era of Emergent Capitalism" (master's thesis, University of California, Berkeley, 1983); Gordon, "Capitalist Development."

26. For a long time, demand models served as the principal alternative to Weberian models, through the work of W. Christaller, *Central Places in Southern Germany* (Englewood Cliffs, 1935; 1966) and A. Lösch, *Die raumliche ordnung der wirtschaft* (Jena, 1944).

27. K. Marx, *Capital* (New York, 1863; 1967); D. Hounshell, *From the American System to Mass Production, 1800–1932* (Baltimore, 1984); B. Page and R. Walker, "From Settlement to Fordism: The Agro-Industrial Revolution in the American Midwest," *Economic Geography* 67 (1991): 281–315; R. Brenner and M. Glick, "The Regulation Approach: Theory and History," *New Left Review* 188 (1991): 45–120.

28. P. Sraffa, "The Laws of Returns under Competitive Conditions," *Economic Journal* 36 (1926): 535–51; N. Kaldor, "The Irrelevance of Equilibrium Economics," *Economic Journal* 82 (1972): 1237–55; R. Walker, "The Dynamics of Value, Price, and Profit," *Capital and Class* 35 (1988): 147–81.

29. G. Allen, *The Industrial Development of Birmingham and the Black Country, 1860–1927* (London, 1929); Haig, *Regional Survey of New York*; M. Wise, "On the Evolution of the Jewelry and Gun Quarters of Birmingham," *Transactions of the Institute of British Geographers* 15 (1949): 57–72.

30. P. David, *Technical Choice, Innovation, and Economic Growth* (New York, 1975); R. Samuel, "Workshop of the World: Steam Power and Hand Technology in Mid-Victorian Britain," *History Workshop* 3 (1977): 6–72; C. Freeman, *The Economics of Industrial Innovation* (London, 1982); R. Walker, "Machinery, Labour, and Location," in *The Transformation of Work?* ed. S. Wood (London, 1989), 59–90; J. Mokyr, *The Lever of Riches: Technological Creativity and Economic Progress* (New York, 1990).

31. Pred, "Intrametropolitan Location"; B. Laurie and M. Schmitz, "Manufacture and Productivity: The Making of an Industrial Base, Philadelphia, 1850–1880," in *Philadelphia: Work, Space, Family, and Group Experience in the Nineteenth Century*, ed. T. Hershberg (New York, 1981), 43–92; P. Scranton, *Proprietary Capitalism: The Textile Manufacture in Philadelphia, 1800–1885* (New York, 1983), and *A Figured Tapestry: Production, Markets, and Power in Philadelphia Textiles, 1885–1941* (New York, 1989); R. Lewis, "Productive and Spatial Structures in the Montreal Tobacco Industry," *Economic Geography* 70 (1994): 370–89.

32. While "industry" is generally taken to mean manufacturing, the same principle applies to all economic activity, including commerce and government, that invests in places and employs people.

33. M. Storper and R. Walker, *The Capitalist Imperative: Territory, Technology, and Industrial Growth* (Cambridge, 1989). Also see A. Scott, *New Industrial Spaces* (London, 1988).

34. On the regional dimensions of the same process, see S. Pollard, *Peaceful Conquest: The Industrialization of Europe, 1760–1970* (New York, 1981); D. Massey, *Spatial Divisions of Labour: Social Structures and the Geography of Production* (London, 1984); M. Marshall, *Long Waves of Regional Development* (London, 1986).

35. R. Walker, "A Requiem for Corporate Geography: New Directions in Industrial Organization, the Production of Place, and Uneven Development," *Geografisker Annaler* 71B (1989): 43–68; A. Sayer and R. Walker, *The New Social Economy* (Oxford, 1992).

36. A. Scott, *New Industrial Spaces* and *Metropolis: From the Division of Labor to Urban Form* (Los Angeles, 1988); O. Williamson, *Markets and Hierarchies* (New York, 1975), and *Economic Organization: Firms, Markets, and Policy Control* (Brighton, 1986); A. Chandler, *Strategy and Structure* (Cambridge, 1962), and *The Visible Hand* (Cambridge, 1977); M. Storper and A. Scott, "The Geographical Foundations and Social Regulation of Flexible Production Complexes," in *Territory and Social Reproduction*, ed. J. Wolch and M. Dear (Boston, 1988), 21–40.

37. G. Becattini, "The Development of Light Industry in Tuscany: An Interpretation," *Economic Notes* 3 (1978): 107–23; M. Piore and C. Sabel, *The Second Industrial Divide* (New York, 1984); C. Sabel and J. Zeitlin, "Historical Alternatives to Mass Production: Politics, Markets, and Technology in Nineteenth-century Industrialization," *Past and Present* 108 (1985): 133–76; A. Scott and M. Storper, eds., *Pathways to Industrialization and Regional Development* (London, 1992). Also see A. Marshall, *Industry and Trade: A Study of Industrial Technique and Business Organization* (London, 1919).

38. A. Scott, *Technopolis: High Technology Industry and Regional Development in Southern California* (Los Angeles, 1994).

39. A. Scott, "The Collective Order of Flexible Production Agglomerations: Lessons for Local Economic Development Policy and Strategic Choice," *Economic Geography* 68 (1992): 219–33; A. Saxenian, *Regional Advantage: Industrial Adaption in Silicon Valley and Route 128* (Cambridge, 1994); M. Storper and R. Salais, *Worlds of Production* (Cambridge, 1997).

40. Scranton, *Proprietary Capitalism* and *Figured Tapestry*; E. Muller and P. Groves, "The Emergence of Industrial Districts in Mid-Nineteenth-Century Baltimore," *Geographical Review* 69 (1979): 159–78; F. Viehe, "Black Gold Suburbs: The Influence of the Extractive Industry on the Suburbanization of Los Angeles, 1890–1930," *Journal of Urban History* 8 (1981): 3–26; G. Gad, "Location Patterns of Manufacturing: Toronto in the Early 1880s," *Urban History Review* 22 (1994): 113–38; R. Lewis, "The Development of an Early Suburban Industrial District: The Montreal Ward of Saint-Ann, 1851–1871," *Urban History Review* 19 (1991): 166–80, and "Restructuring and the Formation of an Industrial District in Montreal's East End, 1850–1914," *Journal of Historical Geography* 20 (1994): 143–57.

41. A. Amin and N. Thrift, "Neo-Marshallian Nodes in Global Networks," *International Journal of Urban and Regional Research* 16 (1992): 571–87; M. Storper, "The Limits to Globalization: Technology Districts and International Trade," *Economic Geography* 68 (1992): 60–93; N. Smith, "Geography, Difference, and the Politics of Scale," in *Postmodernism and the Social Sciences*, ed. J. Doherty, E. Graham, and M. Malek (London, 1992); K. Cox, ed., *Spaces of Globalization: Reasserting the Power of the Local* (New York, 1997).

42. S. DeGeer, "The American Manufacturing Belt," *Geografisker Annaler* 9 (1927): 233–359; B. Thomas, *Migration and Economic Growth: A Study of Great Britain and the Atlantic Economy* (Cambridge, 1973); J. DeVries, *European Urbanization, 1500–1800* (Cambridge, 1984); A. Pred, *Urban Growth and City Systems in the United States, 1840–1860* (Cambridge, 1980); Pollard, *Peaceful Conquest*; D. Meyer, "Midwestern Industrialization and the American Manufacturing Belt in the Nineteenth Century," *Journal of Economic History* 49 (1989): 921–37; W. Cronon, *Nature's Metropolis: Chicago and the Great West* (Chicago, 1991); Page and Walker, "From Settlement to Fordism."

43. Sayer and Walker, *New Social Economy*. In this the classic agglomeration theorists were right, e.g., B. Chinitz, "Contrasts in Agglomeration: New York and Pittsburgh," *American Economic Review, Papers and Proceedings* (1962): 279–89; J. Jacobs, *The Economy of Cities* (New York, 1969). Delimiting the metropolis is not always easy, however. Boston has always been a particularly tricky case, with the many satellite industrial towns of eastern Massachusetts. A useful, if naïve, study is F. Blouin, *The Boston Region, 1810–1850* (Ann Arbor, 1978).

44. Scott, *Metropolis, New Industrial Spaces*, and *Technopolis*; J. Vance, *The Continuing City: Urbanization in Western Civilization* (Baltimore, 1977; 1990); S. Hanson and G. Pratt, *Gender, Work, and Space* (New York, 1995).

45. A. Saxenian, "The Urban Contradictions of Silicon Valley: Regional Growth and the Restructuring of the Semiconductor Industry," *International Journal of Urban and Regional Research* 7 (1983): 237–62; K. Nelson, "Labor Demand, Labor Supply, and the Suburbanization of Low-wage Office Work," in *Work, Production, Territory*, ed. A Scott and M. Storper (Boston, 1986), 149–71; D. Hiebert, "The Social Geography of Toronto in 1931: A Study of Residential Differentiation and Social Structure," *Journal of Historical Geography* 21 (1995): 55–74; Y. Schreuder, "The Impact of Labor Segmentation on the Ethnic Division of Labor and the Immigrant Residential Community," *Journal of Historical Geography* 16 (1990): 402–24.

46. M. Storper and R. Walker, "The Theory of Labor and the Theory of Location," *International Journal of Urban and Regional Research* 7 (1983): 1–41, and *Capitalist Imperative*. Gordon, "Capitalist Development"; S. Buder, *Pullman: An Experiment in Industrial Order and Community Planning, 1880–1930* (New York, 1967); B. Berger, *Working Class Suburb: A Study of Auto Workers in Suburbia* (Berkeley, 1960); E. Greer, *Big Steel: Black Politics and Corporate Power in Gary, Indiana* (New York, 1979); P.-A. Linteau, *The Promoters' City: Building the Industrial Town of Maisonneuve, 1883–1918* (Toronto, 1985); H. Silcox, *A Place to Live and Work: The Henry Disston Saw Works and the Tacony Community of Philadelphia* (University Park, Pa., 1994); B. Pietrykowski, "Fordism at Ford: Spatial Decentralization and Labor Segmentation at the Ford Motor Company, 1920–1950," *Economic Geography* 71 (1995): 383–401.

47. Here one has to look to landscape and architectural studies such as E. Relph, *The Modern Urban Landscape* (London, 1987); M. Conzen, ed., *The Making of the American Landscape* (Boston, 1990); D. Bluestone, *Constructing Chicago* (New Haven, 1991); R. Walker, "Landscape and City Life: Four Ecologies of Residence in the San Francisco Bay Area," *Ecumene* 2 (1995): 33–64.

48. A. Sakolski, *The Great American Land Bubble* (New York, 1932); E. Blackmar, *Manhattan for Rent, 1785–1850* (Ithaca, 1989); M. Davis, *City of Quartz: Excavating the Future in Los Angeles* (London, 1990); M. Doucet and J. Weaver, *Housing the North American City* (Montreal, 1991).

49. Clawson, *Suburban Land Conversion*; Walker, "Suburban Solution," and "A Theory of Suburbanization: Capitalism and the Construction of Urban Space in the United States," in *Urbanization and Urban Planning in Capitalist Societies*, ed. M. Dear and A. Scott (New York, 1981), 383–430.

50. Greer, *Big Steel*.

51. R. Wrigley, "Organized Industrial Districts, with Special Reference to the Chicago Area," *Journal of Land and Public Utility Economics* 23 (1947): 180–98; J. Bredo, *Industrial Estates—Tools for Industrialization* (Glencoe, 1960); Relph, *Modern Urban Land-*

scape; J. Findlay, *Magic Lands: Western Cityscapes and American Culture after 1940* (Berkeley, 1992); A. Pratt, *Uneven Reproduction: Industry, Space, and Society* (London, 1994); C. Ruthheiser, *Imagineering Atlanta: Making Place in the City of Dreams* (New York, 1996).

52. Hoyt, *One Hundred Years*; Burns, *Measuring Business Cycles*; W. Isard, "A Neglected Cycle: The Transport-Building Cycle," *Review of Economics and Statistics* 24 (1942): 149–58; M. Abramowitz, *Evidence of Long Swings in Aggregate Construction since the Civil War* (New York, 1964); Thomas, *Migration and Economic Growth*; M. Gottlieb, *Long Swings in Urban Development* (New York, 1976); J. Whitehand, *The Changing Face of Cities: A Study of Development Cycles and Urban Form* (New York, 1987); R. Harris, *Unplanned Suburbs: Toronto's American Tragedy, 1900–1950* (Baltimore, 1996).

53. Hoyt, *One Hundred Years*.

54. D. Harvey, *The Limits to Capital* (Oxford, 1982) and *The Urbanization of Capital* (Baltimore, 1986). On land speculations of the 1980s, see B. Warf, "Vicious Circle: Financial Markets and Commercial Real Estate in the United States," in *Money, Power, and Space*, ed. S. Corbridge, R. Martin, and N. Thrift (Oxford, 1994), 309–26.

55. For a fine treatment of the complex time-space dynamics of uneven growth, see M. Webber and D. Rigby, *The Golden Age Illusion: Rethinking Postwar Capitalism* (New York, 1996).

56. The best overall treatment of this phenomenon is for Paris, in D. Harvey, *Consciousness and the Urban Experience* (Baltimore, 1986).

57. Teaford, *Unheralded Triumph*; M. Schiesl, *The Politics of Efficiency: Municipal Administration and Reform in America, 1880–1920* (Berkeley, 1977); W. Bean, *Boss Ruef's San Francisco* (Berkeley, 1952); J. Weaver, "'Tomorrow's Metropolis' Revisited: A Critical Assessment of Urban Reform in Canada, 1890–1920," in *The Canadian City: Essays in Urban History*, ed. G. Stelter and A. Artibise (Toronto, 1977), 393–418; H. Platt, *City Building in the New South: The Growth of Public Services in Houston, Texas, 1830–1910* (Philadelphia, 1983); P. Kantor, *The Dependent City Revisited: The Political Economy of Urban Development and Social Policy* (Boulder, 1995); A. Orum, *City Building in America* (Boulder, 1995).

58. Gordon, "Capitalist Development"; Walker, *Suburban Solution* and "Transformation of Urban Structure"; Weaver, *Shaping the Canadian City*; Beauregard, *Voices of Decline*; Groth, *Living Downtown*; Rutherford, *Saving the Canadian City*.

59. S. Buder, *Pullman*; A. Mosher, "'Something Better Than the Best': Industrial Restructuring, George McMurtry, and the Creation of the Model Industrial Town of Vandergrift, Pennsylvania, 1883–1901," *Annals of the Association of American Geographers* 85 (1995): 84–107; M. Crawford, *Building the Workingman's Paradise: The Design of American Company Towns* (New York, 1996).

60. Wrigley, "Organized Industrial Districts."

61. J. Teaford, *City and Suburb: The Political Fragmentation of Metropolitan America, 1850–1970* (Baltimore, 1979); A. Markusen, "Class and Urban Social Expenditure," in Tabb and Sawers, *Marxism and the Metropolis*, 90–112. A. Keating, *Building Chicago: Suburban Developers and the Creation of a Divided Metropolis* (Columbus, 1988); R. Mohl, "Shifting Patterns of American Urban Policy since 1900," in *Urban Policy in Twentieth-Century America*, ed. A. Hirsch and R. Mohl (New Brunswick, 1993), 1–45; D. Beeby, "Industrial Strategy and Manufacturing Growth in Toronto, 1880–1910," *Ontario History* 76 (1984): 199–232; J-P. Collin, "La cité sur mesure: Specialisation sociale de l'espace et autonomie municipale dans la banlieue montréalaise, 1875–1920," *Urban*

History Review 13 (1984): 19–34. Greer, *Big Steel*; Linteau, *Promoters' City*; Orum, *City Building in America*; M. Davis, "Sunshine and the Open Shop," *Antipode* 29 (1998): 356–82.

62. R. Walker, *The Planning Function in Urban Government*, 2d ed. (Chicago, 1950); Schiesl, *Politics of Efficiency*; R. Fischler, "Standards of Development" (Ph.D. diss., University of California, Berkeley, 1993); J. Fairfield, "The Scientific Management of Urban Space: Professional City Planning and the Legacy of Progressive Reform," *Journal of Urban History* 20 (1994): 179–204.

63. Walker, *Planning Function*; Clawson, *Suburban Land Conversion*; M. Weiss, *The Rise of the Community Builders: The American Real Estate Industry and Urban Land Planning* (New York, 1982); S. Chase, "The Process of Suburbanization and the Use of Restrictive Deed Covenants as Private Zoning, Wilmington, Delaware, 1900–1941" (Ph.D. diss., University of Delaware, 1995).

64. Walker, "Suburban Solution"; W. Wilson, *The City Beautiful Movement* (Baltimore, 1989); P. Hall, *Cities of Tomorrow: An Intellectual History of Urban Planning and Design in the Twentieth Century* (Oxford, 1988).

65. D. Hammack, "Comprehensive Planning before the Comprehensive Plan: A New Look at the Nineteenth-Century American City," in *Two Centuries of American Planning*, ed. D. Schaffer (Baltimore, 1988); M. Blackford, *The Lost Dream: Businesspeople and City Planning on the Pacific Coast, 1890–1920* (Columbus, 1993); E. Sandweiss, "Fenced-off Corners and Wider Settings: The Logic of Civic Improvement in Early-Twentieth-Century St. Louis," in *Planning the Twentieth-Century American City*, ed. M. Sies and C. Silver (Baltimore, 1996); G. Hise, *Magnetic Los Angeles: Planning the Twentieth-Century Metropolis* (Los Angeles, 1997); G. Brechin, *Imperial San Francisco: Urban Power, Earthly Ruin* (Berkeley, 1998).

66. On growth coalitions, see J. Logan and H. Molotch, *Urban Fortunes: The Political Economy of Place* (Berkeley, 1986); J. Mollenkopf, *The Contested City* (Princeton, 1983).

67. Warner, *Streetcar Suburbs*.

68. T. Cochran and W. Miller, *The Age of Enterprise: A Social History of Industrial America* (New York, 1961), 153. Thanks to Bill Issel for this quotation.

69. M. Davis, "Los Angeles after the Storm: The Dialectics of Ordinary Disaster," *Antipode* 27 (1995): 221–41.

Chapter Three

Acknowledgments: Research for this study was partially supported by grants from the General Research Board, University of Maryland, College Park. The authors wish to thank David Ward, Michael P. Conzen, Whitman Ridgway, Robert D. Mitchell, and members of the Baltimore History Research Seminar for their constructive criticisms of an earlier draft of this paper, and Gail Leithauser for preparation of the maps.

1. A. Pred, *The Spatial Dynamics of U.S. Urban-Industrial Growth, 1880–1914: Interpretive and Theoretical Essays* (Cambridge, 1966) 16–24; T. Hershberg, H. Cox, and D. Light Jr., "The Journey to Work: An Empirical Investigation of Work, Residence, and Transportation, Philadelphia, 1850 and 1880," (paper presented at the eighty-ninth meeting of the American Historical Association, Chicago, 1974); G. Holt: "Urban Mass Transit History: Where We Have Been and Where We Are Going," in *The National Archives and Urban Research*, ed. J. Finster (Athens, 1974), 81–105.

2. S. B. Warner Jr., *The Urban Wilderness: A History of the American City* (New York, 1972), 62–67.

3. J. Vance, "Housing the Worker: Determinative and Contingent Ties in Nineteenth-Century Birmingham," *Economic Geography* 43 (1967): 125; D. Ward, "Victorian Cities: How Modern?" *Journal of Historical Geography* 1 (1975): 137–38. Mercantile city here includes the term commercial city.

4. Ward, "Victorian Cities," 138.

5. M. Katz, "The People of a Canadian City: 1851–52," *Canadian Historical Review* 53 (1972): 407–8; A. Pred, "Manufacturing in the American Mercantile City: 1800–1840," *Annals of the Association of American Geographers* 56 (1966): 307–88; D. Montgomery, "The Working Classes of the Pre-industrial American City, 1780–1830," *Labor History* 9 (1968): 3–22; A. Kulikoff, "The Progress of Inequality in Revolutionary Boston," *William and Mary Quarterly*, 3rd ser., 28 (1971): 378.

6. R. Fales and L. Moses, "Land-use Theory and the Spatial Structure of the Nineteenth-Century City," *Papers, Regional Science Association* 28 (1972): 66–68.

7. R. Elgie, "The Development of Manufacturing in San Francisco in the Period from 1848 to 1880: An Analysis of Regional Locational Factors and Urban Spatial Structure" (master's thesis, University of California, Berkeley, 1966), 75–78; R. Roberts, "The Changing Patterns in Distribution and Composition of Manufacturing Activity in Hamilton between 1861 and 1921" (master's thesis, McMaster University, 1964), 35–39; Pred, "Manufacturing in the American Mercantile City," 326–30.

8. Elgie, "Development of Manufacturing," 79–80; Pred, "Manufacturing in the American Mercantile City," 326.

9. Clearly the degree and timing of spatial sorting in various North American mercantile cities depended on their sizes and period of organization. M. Bowden, "Growth of the Central Districts in Large Cities," in *The New Urban History: Quantitative Explorations by American Historians*, ed. L. Schnore (Princeton, 1975), 83–88.

10. D. Ward, *Cities and Immigrants* (New York, 1971), 87; A. Krim, "The Irish on the Urban Fringe: Derivation of Residential Space in Mid-Nineteenth-Century Cambridge" (paper presented at the eighth meeting of the Eastern Historical Geography Association, Briarcliff, 1974); Elgie, "Development of Manufacturing," 92–97; Pred, "Manufacturing in the American Mercantile City," 325–30.

11. Warner, *Urban Wilderness*, 65.

12. H. Chudacoff, *The Evolution of American Urban Society* (Englewood Cliffs, 1975) 65; Z. Miller, *The Urbanization of Modern America: A Brief History* (New York, 1973), 44–45; Warner, *Urban Wilderness*, 81–83.

13. Ward, *Cities and Immigrants*, 89–91. Montgomery indicated that some factory and putting-out operations existed in large American cities prior to 1830. Montgomery, "Working Classes."

14. S. B. Warner, *The Private City: Philadelphia in Three Periods of Its Growth* (Philadelphia, 1968), 58.

15. Bowden, "Growth of the Central Cities," 83–88.

16. Ward, *Cities and Immigrants*, 92; Warner, *Private City*, 59.

17. Miller, *Urbanization of Modern America*, 33; P. Temin, "Steam and Waterpower in the Early Nineteenth Century," in *The Reinterpretation of American Economic History*, ed. R. Fogel and S. Engerman (New York, 1971), 236–37.

18. C. Green, "Light Manufactures and the Beginnings of Precision Manufacture before 1861," in *The Growth of the American Economy*, ed. H. Williamson (New York, 1944), 229–49.

19. C. Hoffecker, *Wilmington, Delaware: Portrait of an Industrial City, 1830–1910* (Charlottesville, 1974), 15–35.

20. By developing concomitantly with the initial surge of the city's growth, the metal industries of San Francisco commanded and persisted in a central location near the wharves. Elgie, "Development of Manufacturing," 87–88.

21. Ward, *Cities and Immigrants*, 92–93.

22. Warner, *Private City*, 59.

23. J. Vance, "Housing the Worker: The Employment Linkages as a Force in Urban Structure," *Economic Geography* 42 (1966): 294–325.

24. G. Browne, "Baltimore in the Nation, 1789–1861: A Social Economy in Industrial Revolution (Ph.D. diss., Wayne State University, 1973), 317.

25. *Hunt's Merchants' Magazine* (May 1860): 563–75.

26. Department of the Interior, *Manufactures of the United States in 1860, Eighth Census* (Washington, D.C., 1865), 222.

27. C. Varle, *A Complete View of Baltimore* (Baltimore, 1833). The three federal censuses of manufactures before 1850 (1810, 1820, and 1840) were neither comprehensive nor keyed to the identification and location of individual establishments in cities. Moreover, the McLane Report on Manufactures to the House of Representatives in 1833 did not contain a survey of Maryland or Baltimore. *McLane Report on Manufactures, Documents Relative to Manufactures in the United States* (22d Congress, 1st session, House Document 308, vol. 2, 1833).

28. Varle, *Complete View*, 95–104. The satellite location of industrial communities existed for many years. See also J. Sparks, "Baltimore," *North American Review* 20 (1825): 123; Browne, "Baltimore in the Nation," 70–71.

29. *Niles Weekly Register* 40 (Aug. 1831): 433; Browne, "Baltimore in the Nation," 84.

30. Browne, "Baltimore in the Nation," 187–88; *Niles Weekly Register* 45 (Oct. 1833): 83; Varle, *Complete View*, 84.

31. *Matchett's Baltimore Directory*, 1833, 1835–36, 1837–38 (Baltimore, 1833, 1835–36, 1837–38).

32. Varle, *Complete View*; Levi Woodbury, *Report on Steam Engines* (25th Congress, 3rd Session, House Document 21, 1839), 210–11.

33. For example, the list of shipbuilders had to be compiled from *Matchett's Baltimore Directory, 1833* and from *Niles Weekly Register.* At the same time, a perusal of the city business and population directory indicates that various artisan shops of the central area were undercounted, but this source cannot be used systematically because shop owners (that is, establishments) cannot always be differentiated from shop workers, and residences cannot always be distinguished from workplaces.

34. R. Bernard, "A Portrait of Baltimore in 1800: Economic and Occupational Patterns in an Early American City," *Maryland Historical Magazine* 69 (1974): 341–60.

35. *Matchett's Baltimore Directory*; *Niles Weekly Register* 42 (June 1832): 299; Varle, *Complete View*, 145.

36. E. Muller, "Central District Differentiation in Baltimore, 1833–1860" (paper presented at the seventy-third annual meeting of the Association of American Geographers, Salt Lake City, 1977), 6–10.

37. For example, *Baltimore American and Commercial Advertiser* (1 Jan. 1859); *Manufactures of the United States in 1860, Eighth Census*, manuscript schedule, City and County of Baltimore (Maryland State Library, Annapolis). Data on Baltimore manufacturing in this chapter derived from the manuscript schedule of the census, as well as published reports, present problems. The inclusion of only those establishments with more than $500 annual value of production undoubtedly resulted in an undercount of artisan shops. Moreover, enumerators encountered suspicion from some owners and defined industrial activities inconsistently. For example, slaughtering seems to be mixed in with local retail butchers so that the data on meatpacking are difficult to evaluate. See M. Fishbein, "The Censuses of Manufactures: 1810–1890," *National Archives Accessions* 57 (1963): 11; M. Walsh, "The Value of Mid-Nineteenth-Century Manufacturing Returns: The Printed Census and Manuscript Census Compilations Compared," *Historical Methods Newsletter* 4 (1971): 43–51.

38. For a record of Baltimore's annual shipbuilding activity, see Browne, "Baltimore in the Nation," 499.

39. *Hunt's Merchants' Magazine* (May 1860): 568–69; E. Muller and P. Groves, "The Changing Location of the Clothing Industry: A Link to the Social Geography of Baltimore," *Maryland Historical Magazine* 71 (1976): 405–9.

40. These figures are conservative because of the dependence on the enumerators' designations of industry type. Some establishments listed as merchant tailors and metalwares were, by their size of operation and product, probably more appropriately classified with the ready-made clothing and metalworking industries.

41. Because not all establishments were located, the data of this discussion on location do not always match those of earlier discussions of Baltimore's industry in 1860.

42. *Baltimore American and Commercial Advertiser* (10 Jan. 1860), 1; J. Scharf, *History of Baltimore City and County* (Philadelphia, 1881; Baltimore, 1971), part 1, 207–8.

43. Scharf, *History of Baltimore City*, 422–23.

44. *Baltimore American and Commercial Advertiser* (17 July 1860), 1.

45. J. Garonzik, "Urbanization and the Black Population of Baltimore, 1850–1870" (Ph.D. diss., State University of New York, Stony Brook, 1974), 56.

46. Newspaper accounts suggest that there were some oyster-packing activities along the wharves that the census did not list, possibly because of their seasonal operation. *Baltimore American and Commercial Advertiser* (1 Jan. 1859), 2.

47. Muller, "Central City Differentiation," 6.

48. *Baltimore American and Commercial Advertiser* (31 Aug. 1860), 1.

49. Muller, "Central City Differentiation," 9.

50. J. Dorsey and J. Dilts, *A Guide to Baltimore Architecture* (Cambridge, Md., 1973), 80–81.

51. Ward, *Cities and Immigrants*, 87–91.

52. Scharf, *History of Baltimore City*, 212.

53. Although these shops were established prior to 1833, during the railroad's early years, they were expanded in the 1830s and 1840s in response to the growth and success of the railroad. Scharf, *History of Baltimore City*, 326.

54. The general tendency of an ethnic division of labor has been well documented. See T. Hershberg, M. Katz, S. Blumin, L. Glasco, and C. Griffen, "Occupation and Ethnicity in Five Nineteenth-Century Cities: A Collaborative Inquiry," *Historical Methods Newsletter* 7 (1974): 174–216.

55. Scharf, *History of Baltimore City*, 423; Hershberg, Cox, and Light, "The Journey to Work," 14–15; M. Conzen and K. Conzen, "Geographical Structure in Nineteenth-Century Urban Retailing: Milwaukee, 1836–1890," *Journal of Historical Geography* 5 (1979): 45–66; B. Laurie, "Five Companies and Gangs in Southwark: The 1840's," in *The People of Philadelphia: A History of Ethnic Groups and Lower-Class Life, 1790–1940*, ed. A. Davis and M. Haller (Philadelphia, 1973), 71–87.

56. *Hunt's Merchants' Magazine* (July 1850): 34.

57. For such a pattern in Baltimore, see Garonzik, "Urbanization and the Black Population."

58. Ward, *Victorian Cities*, 131–51.

Chapter Four

Acknowledgments: Thanks to Dick Walker for exceedingly helpful comments on earlier drafts of this work.

1. R. Park, E. W. Burgess, and R. D. McKenzie, *The City* (Chicago, 1925).

2. P. Kellogg, *The Pittsburgh Survey: Findings in Six Volumes* (New York, 1909–14); E. E. Pratt, *Industrial Causes of Congestion in New York City* (New York, 1911).

3. At the end of the twentieth century, a group of Los Angeles-based geographers attempted to make the case for that city being the paradigmatic *fin-de-siècle* American urban area and tellingly borrowed the classic's title: A. J. Scott and E. Soja, eds., *The City: Los Angeles and Urban Theory at the End of the Twentieth Century* (Berkeley, 1997). The significance of the University of Chicago's sociological studies of Chicago for shaping academic and wider public perceptions of American urban conditions continues in contemporary discussions of the urban underclass. See W. J. Wilson, *The Truly Disadvantaged: The Inner City, the Underclass, and Public Policy* (Chicago, 1987), and *When Work Disappears: The World of the New Urban Poor* (Chicago, 1996).

4. The Department of Anthropology and Sociology at the University of Chicago, established in 1892, was the first such department in the United States. During the first three decades of the twentieth century, it was widely recognized as the nation's preeminent department. Discussions of the legacy and mystique of the Chicago School are M. Bulmer, *The Chicago School of Sociology: Institutionalization, Diversity, and the Rise of Sociological Research* (Chicago, 1984); L. Harvey, *Myths of the Chicago School of Sociology* (Aldershot, 1987); A. Cortese, "The Rise, Hegemony, and Decline of the Chicago School of Sociology, 1892–1945," *Social Science Journal* 32 (1995): 235–54. An interesting contemporary analysis of the Chicago School, before it had been identified as such, is T. V. Smith and L. D. White, *Chicago: An Experiment in Social Science Research* (Chicago, 1929). The simultaneous founding and prominence of the University of Chicago Geography Department no doubt helps explain the profound influence on the field of urban studies.

5. R. D. McKenzie, "The Ecological Approach to the Study of the Human Community" in Park, Burgess, and McKenzie, *The City*, 73–74.

6. See M. H. Ebner, *Creating Chicago's North Shore: A Suburban History* (Chicago, 1988).

7. An excellent source is A. D. Keating, *Building Chicago; Suburban Developers and the Creation of a Divided Metropolis* (Columbus, 1988). The historical literature on sub-

urbanization is vast, but key references include S. B. Warner, *Streetcar Suburbs: The Process of Growth in Boston, 1870–1900* (Cambridge, 1962); K. Jackson, *Crabgrass Frontier: The Suburbanization of the United States* (New York, 1985); R. Fishman, *Bourgeois Utopias: The Rise and Fall of Suburbia* (New York, 1987); and J. Stilgoe, *Borderland: Origins of the American Suburb, 1820–1939* (New Haven, 1988).

8. A synopsis of the literature can be found in J. W. Harrington and B. Warf, *Industrial Location: Principles, Practice, and Policy* (New York, 1995).

9. W. N. Mitchell and M. J. Jucius, "Industrial Districts of the Chicago Region and Their Influence on Plant Location," *Journal of Business* 6 (1933): 139.

10. M. Storper and R. Walker, *The Capitalist Imperative: Territory, Technology, and Industrial Growth* (New York, 1989).

11. B. L. Pierce, *A History of Chicago*, 3 vols. (New York, 1937); A. T. Andreas, *History of Chicago*, 3 vols. (Chicago, 1884–86); W. Cronon, *Nature's Metropolis* (New York, 1991).

12. H. Hoyt, *One Hundred Years of Land Values in Chicago* (Chicago, 1933), 47.

13. Ibid., 48.

14. The historic route of the Illinois and Michigan Canal recently achieved the well-deserved status of National Heritage Corridor and, accordingly, has been the subject of extensive historical investigation. An excellent guide to both recent and older studies of the canal and its impact throughout the region is M. P. Conzen and K. J. Carr, eds., *The Illinois and Michigan Canal National Heritage Corridor; A Guide to Its History and Sources* (DeKalb, 1988).

15. J. Fellman, "Pre-building Growth Patterns of Chicago," *Annals of the Association of American Geographers* 47 (1957): 63.

16. Hoyt, *One Hundred Years*, 14.

17. Hoyt, *One Hundred Years*, also states that after its slow start, the north side grew faster than any other section of the city from 1837 until 1840, and that "all warehouses for receiving farmers' produce were located on the north bank of the river" (23). After the Dearborn Street Bridge, heavily traveled by farmers with their wagons, became unsafe and was demolished in 1839, the south bank gained the edge in wholesale trade with farmers. Hoyt goes on to say the north side never recovered its business prestige but instead became, during the 1840s, the most fashionable residential section of Chicago.

18. Keating, *Building Chicago*, 21.

19. S. Buder, *Pullman: An Experiment in Industrial Order and Community Planning, 1880–1930* (New York, 1967). Historians of Pullman seem to prefer to characterize the town's location as a veritable wilderness, ignoring developments in nearby Roseland and, more generally, the land speculation that had already washed over the area. An early romanticized view of Pullman's location is quoted in Buder, *Pullman*, 49.

20. Keating, *Building Chicago*, 22.

21. E. Chamberlin, *Chicago and Its Suburbs* (Chicago, 1894), 359.

22. Names of the investors are listed ibid., 362.

23. Ibid.

24. Hoyt, *One Hundred Years*, 214.

25. Cronon, *Nature's Metropolis*, 209.

26. Keating, *Building Chicago*, 14.

27. D. Pacyga and E. Skerrett, *Chicago: City of Neighborhoods* (Chicago, 1986), 506–7.

28. Other areas of mixed industry and working-class housing established during the nineteenth century on Chicago's periphery were never incorporated into the city but remained vibrant centers. One example is Blue Island in South Chicago, settled during by German immigrants in 1835 and still a self-contained suburb. See H. L. Platt, *The Electric City: Energy and the Growth of the Chicago Area, 1880–1930* (Chicago, 1991), 174–77.

29. Keating, *Building Chicago*, 131.

30. Pacyga and Skerrett, *Chicago*, 510.

31. H. Mayer and R. Wade, *Chicago: Growth of a Metropolis* (Chicago, 1969), 18.

32. Hoyt, *One Hundred Years*, 28–29.

33. G. D. Suttles, *The Man-Made City: The Land Use Confidence Game in Chicago* (Chicago, 1990), 5.

34. Keating, *Building Chicago*, 7.

35. Chamberlin, *Chicago and Its Suburbs*, 353.

36. Keating, *Building Chicago*, 15–16.

37. Ibid., 21.

38. Ibid., 20–21, 25–26.

39. Platt, *The Electric City*, 176–77.

40. Not all planned industrial suburbs achieved success, as is evident in the case of Austinville on Chicago's western edge. See Keating, *Building Chicago*, 22.

41. Chamberlin, *Chicago and Its Suburbs*, 138.

42. An excellent analysis of Chicago's post-fire spatial reorganizations is contained in C. Rosen, *The Limits of Power: Great Fires and the Process of City Growth in America* (Cambridge, 1986). Discussion of the rise of the South Branch district is located at 152–53.

43. Rosen, *The Limits to Power*, 157.

44. Ibid., 152.

45. R. Wrigley, "Organized Industrial Districts, with Special Reference to the Chicago Area," *Journal of Land and Public Utility Economics* 23 (1947): 182.

46. Ibid., 183.

47. Ibid., 182–83.

48. Mitchell and Jucius, "Industrial Districts," 143.

49. The classic view is J. Borchert, "American Metropolitan Evolution," *Annals of the Association of American Geographers* 57 (1967): 301–32. See also M. Conzen, "A Transport Interpretation of the Growth of Urban Regions: An American Example," *Journal of Historical Geography* 1 (1975): 361–82.

50. Symptomatic of this view is Cronon, *Nature's Metropolis*.

51. A. Fishlow, *American Railroads and the Transformation of the Antebellum Economy* (Cambridge, 1965).

52. A. Chandler, *Railroads: The Nation's First Big Business* (New York, 1965).

53. More generally, the balance of power between railroads and industry shifted at the turn of the century as industrialists gained the upper hand vis-à-vis the once seemingly omnipotent railroads. In the Union Stock Yards, for example, the packers rather than the railroads were clearly in control by the end of the century.

54. See G. Hise, Chapter 9 of this volume.

55. E. Riley, *The Development of Chicago and Vicinity as a Manufacturing Center Prior to 1880* (Chicago, 1911), 103.

56. J. B. Appleton, *The Iron and Steel Industry of the Calumet District* (University of Illinois Studies in the Social Sciences, 1925).

57. Ebner, *Creating Chicago's North Shore*, 21.

58. H. Mayer, *The Railway Pattern of Metropolitan Chicago* (Chicago, 1943), 89–90.

59. The area along the two rail routes paralleling the north and south branches of the Chicago River had already begun their industrial development prior to the advent of the railroad.

60. Mayer, *Railway Pattern of Metropolitan Chicago*, 90.

61. Cronon, *Nature's Metropolis*, 175.

62. M. Pudup, "Packers and Reapers, Merchants and Manufacturers: Industrial Structuring and Location in an Era of Emergent Capitalism" (master's thesis, University of California, Berkeley, 1983), 57–58.

63. Cronon, *Nature's Metropolis*, 209–10.

64. J. G. Knapp, "A Review of Union Stock Yards History," *University of Chicago Journal of Business* 2 (1924): 333.

65. Pacyga and Skerrett, *Chicago*, 453; T. Jablonsky, *Pride in the Jungle: Community and Everyday Life in Back of the Yards, Chicago* (Baltimore, 1993), 4.

66. These were the Michigan Central, Michigan Southern, Pittsburgh and Fort Wayne, and the Great Eastern. For a complete discussion of the Union Stock Yards layout, see Knapp, "Review," 336.

67. I. Cutler, *Chicago: Metropolis of the Mid-Continent* (Dubuque, 1982), 75.

68. Accounts differ on this point, but the majority view seems to indicate that Packingtown was not part of the original plan for the Union Stock Yards.

69. At least two syndicate members had also been incorporators of the Union Stock Yard and Transit Company. L. C. Wade, *Chicago's Pride: The Stock Yards, Packingtown, and Environs in the Nineteenth Century* (Urbana, 1987), 65.

70. Cited in Wade, *Chicago's Pride*, 66.

71. Jablonsky, *Pride in the Jungle*, 10.

72. By "vanguard firms" I mean those that had pioneered and/or adopted key industry innovations, notably those involving the slaughter and shipment of fresh beef. See Pudup, "Packers and Reapers."

73. M. Walsh, "Pork Packing as a Leading Edge of Midwestern Industry, 1835–1875," *Agricultural History* 51 (Oct. 1977): 702–17.

74. Jablonsky, *Pride in the Jungle*, 11.

75. R. Slayton, *Back of the Yards: The Making of a Local Democracy* (Chicago, 1986), 21.

76. Pacyga and Skerrett, *Chicago*, 473–74.

77. An excellent discussion of how parks, churches, and other urban amenities were used to spur residential development is contained in D. Bluestone, *Constructing Chicago* (New Haven, 1991).

78. An excellent analysis of the significance and structure of the residential real estate industry is M. Weiss, *The Rise of the Community Builders: The American Real Estate Industry and Urban Land Planning* (New York, 1987).

79. Mayer and Wade, *Chicago*, 24.

80. Suttles, *Man-Made City*, 3. Especially valuable is his stinging critique of the Chicago School's influences on how the city came to be understood.

Chapter Five

Acknowledgments: The author would like to thank Sherry Olson, Richard Walker, Ted Muller, Greg Hise, David Meyer, and two referees for their comments, and the Social Sciences and Humanities Research Committee for financial assistance.

1. K. Jackson, *Crabgrass Frontier: The Suburbanization of the United States* (New York, 1985); R. Fishman, *Bourgeois Utopias: The Rise and Fall of Suburbia* (New York, 1987); J. Stilgoe, *Borderland: Origins of the American Suburb, 1820–1939* (New Haven, 1988); S. B. Warner, *Streetcar Suburbs: The Process of Growth in Boston, 1870–1900* (Cambridge, 1962); D. Ward, *Cities and Immigrants* (New York, 1971).

2. H. Douglass, *The Suburban Trend* (New York, 1925); C. Golab, *Immigrant Destinations* (Philadelphia, 1977); R. Harris, "A Working-Class Suburb for Immigrant Toronto, 1909–1913," *Geographical Review* 81 (1991): 318–32; K. Kane and T. Bell, "Suburbs for a Labor Elite," *Geographical Review* 75 (1985): 319–34; R. Lewis, "The Segregated City: Class Residential Patterns and the Development of Industrial Districts in Montreal, 1861 and 1901," *Journal of Urban History* 19 (1991): 123–52; H. McKiven, *Iron and Steel: Class, Race, and Community in Birmingham, Alabama, 1875–1920* (Chapel Hill, 1995); G. Taylor, *Satellite Cities: A Study of Industrial Suburbs* (New York, 1915).

3. Jackson, *Crabgrass Frontier*; A. Pred, "The Intrametropolitan Location of American Manufacturing," *Annals of the Association of American Geographers* 54 (1964): 165–89; A. Scott, "Locational Patterns and Dynamics of Industrial Activity in the Modern Metropolis," *Urban Studies* 19 (1982): 111–42; Ward, *Cities and Immigrants*.

4. S. Buder, *Pullman: An Experiment in Industrial Order and Community Planning, 1880–1930* (New York, 1967); G. Gad, "Locational Patterns of Manufacturing: Toronto in the Early 1880s," *Urban History Review* 22 (1994): 113–38; R. Lewis, "The Development of an Early Suburban Industrial District: The Montreal Ward of Saint-Ann, 1851–1871," *Urban History Review* 19 (1991): 166–80, and "Restructuring and the Formation of an Industrial District in Montreal's East End, 1850–1914," *Journal of Historical Geography* 20 (1994): 143–57; E. Muller and P. Groves, "The Emergence of Industrial Districts in Mid-Nineteenth-Century Baltimore," *Geographical Review* 69 (1979): 159–78; F. Viehe, "Black Gold Suburbs: The Influence of the Extractive Industry on the Suburbanization of Los Angeles, 1890–1930," *Journal of Urban History* 8 (1981): 3–26.

5. In this chapter, "suburban" refers to both independent municipalities adjacent to the city and to the newly built-up districts within the city boundary. While political jurisdiction did have some important implications for differences in the character of city and independent suburban development, the similarities in the formation of industrial districts on the urban fringe (city and non-city) were striking.

6. Beginning in 1847 the assessments provide an annual rent valuation for every residence and business in the city, and, at later dates, for surrounding municipalities. The rents of all manufacturing firms in the city and the adjacent towns for 1861, 1890, and 1929 were collected and analyzed. Given that the published manufacturing census does not provide firm-specific data and that, with the exception of 1871, the manuscript censuses were either destroyed or are publicly inaccessible, the assessments are the best source available for constructing the location, character, and establishment of Montreal's industrial districts. For more detail on the assessment rolls and the way they

have been used to determine the social and industrial geography of Montreal, see R. Lewis, *Manufacturing Montreal: The Making of an Industrial Landscape, 1580 to 1930* (Baltimore, 2000), 273–78; D. Hanna and S. Olson, "Métiers, loyers et bouts de rue: L'armature de la société montréalaise de 1881 á 1901," *Cahiers de Géographie du Québec* 27 (1983): 255–75.

7. D. Gregory, "'A New and Differing Face in Many Places': Three Geographies of Industrialization," in *A Historical Geography of England and Wales*, ed. R. Dodgson and R. Butlin (London, 1990), 351–99; Lewis, "Productive and Spatial Strategies"; W. Licht, *Getting Work: Philadelphia, 1840–1950* (Cambridge, 1992); P. Scranton, *Proprietary Capitalism: The Textile Manufacture at Philadelphia, 1800–1885* (New York, 1983); R. Walker, "The Geographical Organization of Production Systems," *Environment and Planning D* 6 (1988): 377–408.

8. The classic example of this is the Chicago district of Pullman. See Buder, *Pullman*.

9. This is not to say that the conflicts between different alliances within the city were not important in the shaping of the city. They often were. For example, the opportunities for suburban development were greatly restricted by the Montreal Street Railway Company's nonexpansion policy. Pitted against land developers, commuters, and city and suburban councils, the company viewed the demands for wholesale expansion as a shaky financial proposition and resisted all demands for the extension of tracks into areas with low population density. The point, however, is that despite the internal differences, local alliances more often had the same end in sight—advancing both their and the city's economic success.

10. For more detail on the rise and fall of Old Montreal as a manufacturing district in the nineteenth century, see R. Lewis, "Productive Strategies and Manufacturing Reorganization in Montreal's Central District, 1850–1900," *Urban Geography* 16 (1995): 4–22.

11. P. Bischoff, "Des forges du Saint-Maurice aux fonderies de Montréal: mobilité géographique, solidarité communautaire et action syndicale les mouleurs, 1829–1881," *Revue d'histoire de l'Amerique française* 43 (1989): 3–29; J. Burgess, "L'industrie de la chassure à Montréal: 1840–1870—Le passage de l'artisanat à la fabrique," *Revue d'histoire de l'Amerique française* 31 (1977): 187–210; M. Denison, *The Barley and the Stream: The Molson Story* (Toronto, 1955); J. Ferland, "Evolution des rapports sociaux dans l'industrie canadienne du cuir au tournant du 20e siècle" (Ph.D. diss., McGill University, 1985); J. Hamelin and Y. Roby, *Histoire économique du Québec, 1851–1896* (Montreal, 1971); F. Harvey, *Révolution industrielle et travailleurs* (Montreal, 1978); G. Teal, "The Organization of Production and the Heterogeneity of the Working Class: Occupation, Gender, and Ethnicity among Clothing Workers in Quebec" (Ph.D. diss., McGill University, 1986).

12. E. A. Cureton, "The Lachine Canal" (master's thesis, McGill University, 1957); Lewis, "Development of an Early Suburban Industrial District"; L. McNally, *Waterpower on the Lachine Canal, 1846–1900* (Ottawa, 1982); G. Tulchinsky, *The River Barons: Montreal Businessmen and the Growth of Industry and Transportation, 1837–1853* (Toronto, 1977); J. Willis, *The Process of Hydraulic Industrialization on the Lachine Canal: Origins, Rise, and Fall* (Ottawa, 1987).

13. Cureton, "Lachine Canal"; La Société historique de Saint-Henri, *Portrait d'une ville: Saint-Henri 1875–1905* (Montreal, 1987); G. Lauzon and L. Ruellard, *1875/Saint-Henri* (Montreal, 1985).

14. Anon., "Our Canadian Industries II. The Canadian Rubber Co., of Montreal," *Dominion Illustrated* (7 Dec. 1889), 359; Atelier d'histoire Hochelaga-Maisonneuve, *L'industrialisation à Hochelaga-Maisonneuve 1900–1930* (Montreal, 1980), and *De fil en aiguille: chronique ouvière d'une filature de coton à Hochelaga en 1880* (Montreal, 1985); Lewis, "Restructuring and the Formation of an Industrial District."

15. The creation of working-class districts in the inner city, stratified by ethnicity, and the growth of middle-class suburbs has been well documented for North American cities. The importance of working-class suburbs and their relationship to industrial suburbanization, however, has been downplayed. For the development of a more complex social geography in Montreal, see Atelier, D'histoire, *De fil en aiguille*; B. Bradbury, *Working Families: Age, Gender, and Daily Survival in Industrializing Montreal* (Toronto, 1993); D. Hanna, "Montreal: A City Built by Small Builders, 1867–1896 (Ph.D. diss., McGill University, 1986); Lauzon and Ruellard, *1875/Saint-Henri*; Lewis, "Segregated City." One example of a predominantly centralized immigrant population were Jews who were to be found close to the expanding inner-city garment industry. See J. Siedel, "The Development and Social Adjustment of the Jewish Community in Montreal (master's thesis, McGill University, 1939). For the development of middle-class suburbs, see J-P. Collin, "La cité sur mesure: Spécialisation sociale de l'espace et autonomie municipale dans la banlieue montréalaise, 1875–1920," *Urban History Review* 13 (1984): 19–34; W. van Nus, "The Role of the Suburban Government in the City-Building Process: The Case of Notre-Dame-de-Grace, Quebec, 1876–1910," *Urban History Review* 13 (1984): 91–103. Examples of other cities include Golab, *Immigrant Destinations*; McKiven, *Iron and Steel.*

16. R. Hoskins, "An Analysis of the Payrolls of the Point St. Charles Shops of the Grand Trunk Railway," *Cahiers de Géographie du Québec* 33 (1989): 323–44; P. Thornton, S. Olson, and Q. Thuy Thach, "Dimensions sociales de la mortalité infantile à Montréal au milieu XIXe siècle," *Annales de Démogaphie Historique* (1988): 299–325.

17. Atelier, d'Histoire Hochelaga-Maisonneuve, *L'Histoire du logement ouvrier à Hochelaga-Maisonneuve* (Montreal, 1980); M. Choko, *Crises du logement à Montréal (1860–1939)* (Montreal, 1980); M. Choko and R. Harris, "The Local Culture of Property: A Comparative History of Housing Tenure in Montreal and Toronto," *Annals of the Association of American Geographers* 80 (1990): 73–95; T. Copp, *The Anatomy of Poverty: The Condition of the Working Class in Montreal, 1897–1929* (Toronto, 1974); Hanna, "Montreal"; S. Hertzog and R. Lewis, "A City of Tenants: Homeownership and Social Class in Montreal, 1847–1881," *Canadian Geographer* 30 (1986): 316–23; G. Lauzon, *Habitat ouvrier et révolution industrielle: Le cas du village St-Augustin* (Montreal, 1989).

18. Lewis, "Industry and Space," 286–91; McNally, *Water Power on the Lachine Canal*; Willis, *Process of Hydraulic Industrialization*; B. Young, *In Its Corporate Capacity: The Seminary of Montreal as a Business Institution, 1816–1876* (Montreal, 1986).

19. As in most other cities, investment in housing tended to lag behind other sectors because housing was seen as risky, turnover of capital was slow, and most house builders could not compete in money markets with commercial and industrial investors.

20. The career of George-Etienne Cartier neatly illustrates the intersection of local, provincial, and federal politics, pursuit of private gain through land ownership and company directorships, and the cultural networks and ideological framework of

the Montreal elite. See B. Young, *George-Etienne Cartier: Montreal Bourgeois* (Montreal, 1981).

21. Even though the Harbour Commission was dominated by Anglophones, French-Canadian members of Montreal's bourgeoisie with interests in railways, banks, and utilities were active after 1850. See Brouillard, "La Commision du Havre de Montréal (1850–1896)," in Société historique de Montréal, *Montréal: Artisans, histoire, patrimoine* (Montreal, 1979), 83–102.

22. B. Brouillette, "Les ports et les transports," in *Montréal economique*, ed. E. Minville (Montreal, 1943), 114–82; Hamelin and Roby, *Histoire économique*.

23. P. Craven and T. Traves, "Canadian Railways as Manufacturers, 1850–1880," Canadian Historical Association, *Historical Papers* (1983): 254–81.

24. Anon, "Ville St. Pierre, P.Q.," *Canadian Municipal Journal* 11 (Jan. 1915): 27–33; Cureton, *Lachine Canal*; R. Lewis, "Productive and Spatial Strategies in the Montreal Tobacco Industry, 1850–1918," *Economic Geography* 70 (1994): 370–89; Delage, "L'industrie manufacturière," 183–241; La Société historique de Saint-Henri, *Portrait d'une ville*.

25. Atelier, D'histoire, *L'industrialisation*; Delage, "L'industrie manufacturière"; P.-A. Linteau, *The Promoters' City: Building the Industrial Town of Maisonneuve, 1883–1918* (Toronto, 1985).

26. The quotation is from M. Gauvin, "The Reformer and the Machine: Montreal Civic Politics from Raymond Préfontaine to Méderic Martin," *Journal of Canadian Studies* 13 (1978): 23. Also see G. Bourassa, "Les élites politiques de Montréal: De l'aristocratie à la démocratie," *Canadian Journal of Economics and Political Science* 31 (1965): 35–51; A. Germain, *Les Mouvements de réforme urbaine à Montréal au tournant du siècle* (Montreal, 1984); A. Gordon, "Ward Heelers and Honest Men: Urban Québécois Political Culture and the Montreal Reform of 1909," *Urban History Review* 23 (1995): 20–32.

27. D. Cross, "The Neglected Majority: The Changing Role of Women in Nineteenth-Century Montreal," in *The Canadian City*, ed. G. Stelter and A. Artibise (Toronto, 1977), 255–81. For a discussion of female employment in general, see Bradbury, *Working Families*, and for the suburbs see Lewis, "Restructuring and the Formation of an Industrial District," and "Productive and Spatial Strategies."

28. R. Binns, *Montreal's Electric Streetcars: An Illustrated History of the Tramway Era, 1892 to 1959* (Montreal, 1973).

29. For a discussion of Montreal's mass transit system, see C. Armstong and H. Nelles, "Suburban Street Railway Strategies in Montreal, Toronto, and Vancouver, 1869–1930," in *Power and Place: Canadian Urban Development in the North American Context*, ed. G. Stelter and A. Artibise (Vancouver, 1986), 187–218, and *Monopoly's Moment: The Organization and Regulation of Canadian Utilities, 1830–1930* (Philadelphia, 1986) 34–55, 248–54; Linteau, *Promoters' City*, 87–96.

30. The quotation is from B. Ramirez and M. Del Balso, "The Italians of Montreal: From Sojourning to Settlement, 1900–1921," in *Little Italies in North America*, ed. R. Harney and J. Scarpaci (Toronto, 1981) 75; J. Boissevain, *The Italians of Montreal* (Ottawa, 1970), 1–4; B. Ramirez, "Montreal's Italians and the Socioeconomy of Settlement: Some Historical Hypotheses," *Urban History Review* 10 (1981): 39–48.

31. C. Young, *The Ukrainian Canadians* (Toronto, 1931), 114–25.

32. E. Chambers, *Suburban Montreal as Seen from the Routes of the Park and Island Railway Co.* (Montreal, 1895); Collin, "La cité sur mesure"; Hanna, "Montreal"; G. Levine, "Class, Ethnicity, and Property Transfers in Montreal, 1907–1909," *Journal of Historical Geography* 14 (1988): 360–80; Linteau, *Promoters' City*; Ramirez, "Montreal's Italians"; Société historique de Saint-Henri, *Portrait d'une ville*; van Nus, "Role of the Suburban Government."

33. Collin, "Cité sur mesure"; Linteau, *Promoters' City*.

34. R. Blanchard, *L'ouest du Canada français: Montréal et sa region* (Montreal, 1953), 279–80; M. Casey, "The Use of Power for Port Facilities," *Engineering Journal* 7 (1924): 486–88; F. Cowie, "The Great National Port of Canada: Features of the Important Extension Work in Progress in Montreal Harbour," *Canadian Engineer* (1912): 178–83; P.-A. Linteau, "Le dèveloppement du port de Montréal au dèbut du 20e siècle," Canadian Historical Association, *Historical Papers* (1972): 181–205; G. Taylor, "A Merchant of Death in the Peaceable Kingdom: Canadian Vickers, 1911–1927," in *Canadian Papers in Business History*, ed. P. Baskerville (Victoria, 1989), 1:213–44. It should be noted that the use of trucks appears to have been minimal in this period. In his interviews with company executives of firms located on the Lachine Canal, Cureton found that the transportation of goods by water was far more important than transportation by truck until the middle of the 1930s. Cureton, "Lachine Canal."

35. R. Naheut, "Une expérience canadienne de taylorisme: Le cas des usines Angus du Candien Pacifique" (Maitrise, L'Université du Québec à Montréal 1984).

36. J. Dales, *Hydroelectricity and Industrial Development: Quebec, 1898–1940* (Cambridge, 1957); C. Hogue, A. Bolduc, and D. Larouche, *Québec: Un siècle d'electricité* (Montreal, 1979); M. Martin, "Communication and Social Forms: The Development of the Telephone, 1876–1920," *Antipode* 23 (1991): 307–33.

Chapter Six

Acknowledgments: Thanks to Kevin Carew and David Landau for research assistance, and to Gray Brechin, Jim Buckley, Paul Groth, and Paul Rhode for advice and counsel.

1. For the traditional view, see J. Vance, *Geography and Urban Evolution in the San Francisco Bay Area* (Berkeley, 1964), 50–51; P. Groves, *The Intrametropolitan Location of Manufacturing in the San Francisco Bay Area* (Ph.D. diss., University of California, Berkeley, 1969), 28–29.

2. See R. Walker and R. Lewis, Chapter 2 of this volume.

3. C. McWilliams, *California: The Great Exception* (New York, 1949); G. Barth, *Instant Cities: Urbanization and the Rise Of San Francisco and Denver* (New York, 1975); L. Doti and L. Schweikart, *Banking in the American West: From the Gold Rush to Deregulation* (Norman, 1991); G. Brechin, *Imperial San Francisco: Urban Power, Earthly Ruin* (Berkeley, 1998).

4. On early resource extraction and processing around the bay, see J. Hittell, *The Commerce and Industries of the Pacific Coast* (San Francisco, 1882); A. Hynding, *From Frontier to Suburb: The Story of the San Mateo Peninsula* (Belmont, 1981); M. Koch, *Santa Cruz County: Parade of the Past* (Santa Cruz, 1973); R. Paul, "The Wheat Trade between California and the United Kingdom," *Mississippi Valley Historical Review* 45 (1973): 391–412; J. Hutchinson, "Northern California from Haraszthy to the Beginnings of

Prohibition," in *The Book of California Wine*, ed. D. Muscatine, M. Amerine, and B. Thompson (Berkeley, 1984), 30–48.

5. M. Gordon, *Employment Expansion and Population Growth* (Berkeley, 1954); J. Guinn, *History of the State Of California and Biographical Record of Oakland and Environs* (Los Angeles, 1907); Vance, *Geography and Urban Evolution*.

6. W. Issel and R. Cherny, *San Francisco, 1865–1932* (Berkeley, 1986), 25, 54; R. Elgie, "The Development of San Francisco Manufacturing, 1848–1880" (master's thesis, University of California, Berkeley, 1966); N. Shumsky, *Tar Flat and Nob Hill: A Social History of Industrial San Francisco during the 1870s* (Ph.D. diss., University of California, Berkeley, 1972), 22–24; R. Walker, "Another Round of Globalization in San Francisco," *Urban Geography* 17 (1996): 60–94.

7. Keeping in mind that Currier and Ives's birds'-eye views were sometimes embellished.

8. On capital, see R. Trusk, "Sources of Capital of Early California Manufacturers, 1850–1880" (Ph.D. diss., University of Illinois, 1960); Issel and Cherny, *San Francisco*; Brechin, *Imperial San Francisco*. On the stimulus of labor migration, see Elgie "San Francisco Manufacturing." On the skewed occupational structure toward professions and craft skills, see Issel and Cherny, *San Francisco*, 54–55. Wages, incomes, and value added per worker were all higher in California and the West than elsewhere in the United States. Average firm size was lower. For evidence of innovation, see Hittell, *Commerce and Industries*, Brechin, *Imperial San Francisco*; J. Johnson, "Early Engineering Center in California," *California Historical Quarterly* 29 (1950): 193–209.

9. For an overview of California's resource-based economic dynamism at this time, see R. Walker, "California's Golden Road to Riches," *Annals of the Association of American Geographers* 91 (2001): 167–99.

10. I consider animal processing to be a more useful category than simply meatpacking, because it includes byproducts, like glue, and marine products, like whale baleen and fish rendering.

11. The following portrait is a composite over thirty years, which means companies came and went and may not always be strictly contemporaneous. It is drawn from my close readings of the Sanborn Insurance Maps for 1886–93 and 1899–1905, and from F. Hackett, ed., *The Industries of San Francisco* (San Francisco, 1884), Anon., *The Bay of San Francisco: The Metropolis of the Pacific Coast and Its Suburban Cities* (Chicago, 1892); Hittell, *Commerce and Industries*; Trusk, "Sources of Capital"; Elgie "San Francisco Manufacturing"; Shumsky, *Tar Flat and Nob Hill*; Issel and Cherny, *San Francisco*.

12. On San Francisco's emergent and shifting business district, see M. Bowden, "The Dynamics of City Growth: An Historical Geography of the San Francisco Central District, 1850–1931" (Ph.D. diss., University of California, Berkeley, 1967). I have also become aware of a significant separation of company offices and salesrooms from manufacturing well before 1900 but cannot pursue it here.

13. For the location of businesses north of Market see San Francisco Post Company, *Business Map* (San Francisco, 1880). On Chinese labor, see A. Saxton, *The Indispensable Enemy: Labor and the Anti-Chinese Movement in California* (Berkeley, 1971). On publishing, see J. Bruce, *Gaudy Century: The Story of San Francisco's Hundred Years of Robust Journalism* (New York, 1948). Figures on printers from G. Bowser, *A Business Directory of the City and County of San Francisco* (San Francisco, 1885).

14. On machining and mining, see J. Blum, "The Early San Francisco Iron Industry" (manuscript, 1997); Vance, *Geography and Urban Evolution*; Brechin, *Imperial San Francisco*.

15. On the lumber industry, see J. Buckley, "Building the Redwood Region: The Redwood Lumber Industry and the Landscape of Northern California, 1850–1929" (Ph.D. diss., University of California, 2000). Mission Bay was filled and the wood district fully in place by 1884. N. Olmsted, *Vanished Waters: A History of San Francisco's Mission Bay* (San Francisco, 1986).

16. On Potrero Point, see Olmsted, *Vanished Waters*; P. Groth, "Making the System Work: Engineering Cultural Change in 20th-Century Factory Complexes" (paper given at the American Studies Association meeting, New York, 1988); J. Kemble, *San Francisco Bay: A Pictorial Maritime History* (Cambridge, Md., 1957), 10 (map).

17. Butchertown's history has not been told, but on Miller and Lux, see D. Igler, *Industrial Cowboys: Nature, Private Property, and the Regional Expansion of Miller & Lux, 1850–1920* (Berkeley, 2001).

18. On residential patterns, their linkage to industrial expansion, and bourgeois resettlement, see Shumsky, *Tar Flats and Nob Hill*, 138–39; Issel and Cherny, *San Francisco*; Buckley, "Building the Redwood Region"; R. Dillon, *North Beach: The Italian Heart of San Francisco* (Novato, 1985); A. Shumate, *Rincon Hill and South Park: San Francisco's Early Fashionable Neighborhood* (Sausalito, 1988); P. Groth, *Living Downtown: The History of Residential Hotels in The United States* (Berkeley, 1994); R. Walker, "Landscape and City Life: Four Ecologies of Residence in the San Francisco Bay Area," *Ecumene* 2 (1995): 33–64.

19. On Dogtown, see Olmsted, *Vanished Waters*, 47; Groth, "Making the System Work." On land speculation by major San Francisco capitalists, see Brechin, *Imperial San Francisco*. There is no detailed study of land speculation and residential development in the southeastern part of San Francisco, however.

20. L. Kauffman, *South San Francisco: A History* (self-published, 1976); J. Blum, "South San Francisco: The Making of an Industrial City," *California History* 63 (1984): 119; Igler, *Industrial Cowboys*.

21. E. Burns, "The Process of Suburban Residential Development: The San Francisco Peninsula, 1860–1970" (Ph.D. diss., University of California, Berkeley, 1975); Walker, "Landscape and City Life"; Kauffman, *South San Francisco*; Hynding, *From Frontier to Suburb*. Kauffman and Hynding are descendants of South City leaders.

22. Hynding, *From Frontier to Suburb*; Blum, "Early San Francisco Iron Industry." Although the Peninsula was chiefly a commuter zone for San Francisco, an electronics industry grew up there early in the century. It developed out of high-voltage transmission and long-distance communications, and the vacuum tube invented in Palo Alto by Lee DeForest. D. Hanson, *The New Alchemists: Silicon Valley and the Microelectronics Revolution* (Boston, 1982); T. Sturgeon, "The Origins of Silicon Valley: The Development of the Electronics Industry in the San Francisco Bay Area" (master's thesis, University of California, Berkeley, 1992). The airport and its surrounding warehousing district did not arrive until World War II.

23. On the importance of San Francisco's coffee companies nationally, see M. Pendergrast, *Uncommon Grounds: The History of Coffee and How It Transformed the World* (New York, 1999).

24. P. Cohen, "Transformation in an Industrial Landscape: San Francisco's Northeast Mission" (master's thesis, San Francisco State University, 1998). On industrial conditions in San Francisco after World War I, see U.S. Bureau of the Census, *Fourteenth Census of the United States, State Compendium—California* (Washington, D.C., 1924) 158–59; San Francisco Chamber of Commerce, *Directory of Manufactures of San Francisco, California* (San Francisco, 1920, 1922); San Francisco Chamber of Commerce, *San Francisco Economic Survey* (San Francisco, 1937–40).

25. Los Angeles outgrew the Bay Area in population and employment 1880–1910 and 1920–40. On San Francisco's declining commercial dominance over the West, see E. Pomeroy, *The Pacific Slope* (New York, 1965); on its continuing financial hegemony, see Doti and Schweikert, *Banking in the American West*. Figures on shipping are from R. Calkins and W. Hoadley, *An Economic and Industrial Survey of the San Francisco Bay Area* (Sacramento, 1941), 156–58.

26. Calkins and Hoadley, *Economic and Industrial Survey*, 170. A few local historians have alerted us to the need not to confuse San Francisco's performance with that of the whole Bay Area, e.g., M. Scott, *The San Francisco Bay Area: A Metropolis in Perspective* (Berkeley, 1959), 136; Issel and Cherny, *San Francisco*, 50. Industrial employment in the Bay Area fell in the period 1904–9 and grew little up to 1914, according to census figures, although output continued to expand; however, the census defined the metro area to exclude much of Contra Costa. U.S. Bureau of the Census, *Fourteenth Census*, 27. The evidence never bore out the corporate takeover theory, as shown fifty years ago by A. Trice, "California Manufacturing Branches of National Firms, 1899–1948: Their Place in the Economic Development of the State" (Ph.D. diss., University of California, Berkeley, 1955). On Studebaker, see T. Bonsall, *More Than They Promised: The Studebaker Story* (Stanford, 2000). An aside: even without national corporations, generic names for industrial companies did not begin with General Foods and General Motors; San Francisco's early capitalists were remarkably uninspired in their naming, repeating over and over the same handful of labels: Pioneer, Western, Pacific Coast, California, San Francisco, Enterprise, etc.

28. Quotation from Grant in Issel and Cherny, *San Francisco*, 51. Gerstle quotation from President's Annual Report, Sixty-First Annual report of the Chamber of Commerce of San Francisco, 17–18, as cited in Scott, *San Francisco Bay Area*. On San Francisco's militant labor in general, see McWilliams, *California*; M. Kazin, *Barons of Labor: The San Francisco Building Trades and Union Power in The Progressive Era* (Urbana, 1987). On the virtuous circle of high wages and growth, see Gordon, *Employment Expansion*. On the Chinese, see Anon., *Bay of San Francisco*; Saxton, *Indispensable Enemy*. Many Chinese migrated to southern California in the 1880s and 1890s to escape persecution, no doubt helping manufacturing growth in Los Angeles.

29. On the southern California capitalists' militant self-organizing in the Progressive Era, see M. Davis, *City of Quartz* (London, 1989); W. Deverell and T. Sitton, eds., *California Progressivism Revisited* (Berkeley, 1994).

30. On San Francisco's business power and alliances, see McWilliams, *California*; Issel and Cherny, *San Francisco*; Brechin, *Imperial San Francisco*; Scott, *San Francisco Bay*; J. Kuhn, *Imperial San Francisco: Politics and Planning in an American City, 1897–1906* (Lincoln, 1979); W. Issel, "Citizens Outside the Government: Business and Urban Policy in San Francisco and Los Angeles, 1890–1932," *Pacific Historical Review* 57 (1988): 117–45; W. Issel, "Business Power and Political Culture in San Francisco,

1900–1940," *Journal of Urban History* 16 (1989): 52–77; W. Issel, "The New Deal and Wartime Origins of San Francisco's Postwar Political Culture: The Case of Growth Politics and Policy," in *The Way We Really Were: The Golden State in the Second World War,* ed. R. Lotchin (Urbana, 2000), 68–92.

31. On geographical industrialization, see M. Storper and R. Walker, *The Capitalist Imperative: Territory, Technology, and Industrial Growth* (New York, 1989).

32. For an overview of change in California industry, see Gordon, *Employment Expansion.* I address the changes in specific sectors later in the chapter.

33. The chief source on industrial Oakland, including locations, is E. Hinkel and W. McCann, eds., *Oakland, 1852–1938* (Oakland, 1939), chap. 12. See also Illustrated Directory Company, *The Illustrated Directory Of Oakland, California* (Oakland, 1896); Anon., *Greater Oakland, 1911* (Oakland, 1911); Oakland Central National Bank, *Oakland, California: The City of Diversified Industry* (Oakland, 1920); Oakland Tribune, *Year Book* (Oakland, 1926, 1927); R. Cleland and O. Hardy, *March of Industry* (Los Angeles, 1929); Oakland Chamber of Commerce, *Industrial Facts about Oakland and Alameda County, California* (Oakland, 1931); Emeryville Industries Association, *Emeryville, California: Facts and Factories* (Emeryville, 1935), Emeryville Industries Association, *A Roster of Emeryville Industries* (Emeryville, c. 1936).

34. Carpentier owned the waterfront from 1852 to 1868, then ceded it to the Oakland Waterfront Company, held by himself, his brother, and Samuel Merritt, as well as Leland Stanford and other San Francisco barons. J. Dykstra, "A History of the Physical Development of the City of Oakland: The Formative Years, 1850–1930" (master's thesis, University of California, Berkeley, 1967); B. Bagwell, *Oakland: Story of a City* (Novato, 1982).

35. Calkins and Hoadley, *Economic and Industrial Survey,* 217, 156, and 212. Hinkel and McCann, *Oakland,* call Oakland "the Glasgow of the US, the Marseilles of the Pacific and the Detroit of the West," boosterist terms promoted by local business leaders in the preceding years. Published figures can be misleading because much of East Bay industry was in unincorporated areas or because of exaggeration by Chamber of Commerce–type sources.

36. U.S. Bureau of Census, *Census of Manufactures, 1914,* I, 179. There were seventeen CalPak canneries in the south county alone. Oakland Chamber of Commerce, *Industrial Facts,* 22. On Bay Area canning, see J. Cardellino, "Industrial Location: A Case Study of the California Fruit and Vegetable Canning Industry, 1860 to 1984" (master's thesis, University of California, Berkeley, 1984); W. Braznell, *California's Finest: The History of Del Monte Corporation and the Del Monte Brand* (San Francisco, 1982); Hackett, *Industries of San Francisco;* Hinkel and McCann, *Oakland;* Calkins and Hoadley, *Economic and Industrial Survey.*

37. Hinkel and McCann, *Oakland;* Emeryville Industries Association, *Emeryville.* Overall, California's steel and machinery industries made spectacular advances in the 1910s, due in part to low-cost energy and federal wartime spending. Gordon, *Employment Expansion,* 56.

38. Hinkel and McCann, *Oakland;* J. Moore, *The Story of Moore Dry Dock Company: A Picture History* (Sausalito, 1994).

39. Hinkel and McCann, *Oakland;* Oakland Tribune, *Year Book 1926,* 53, 181; H. Christman, "Development of the Pacific Coast Automotive Industry," *Western Machinery World* (Jan. 1929): 13–19.

40. On the residential expansion of Oakland, see Dykstra, *History of Oakland*, and Bagwell, *Oakland*. On local real estate cycles, see L. Maverick, "Cycles in Real Estate Activity," *Journal of Land and Public Utility Economics* 8 (1932): 191–99. The Realty Syndicate was responsible for about half of modern Oakland, especially along the foothills, where middle- and upper-class riders used the trolleys to commute from homes in the elite districts such as Claremont, Elmwood, Piedmont, and Trestle Glen. See also Walker, "Landscape and City Life."

41. Emeryville Industries Association, *Emeryville*. Emeryville's political history has not been adequately told. Emeryville preceded the first industrial suburb of Los Angeles, Vernon, by eight years. On Berkeley zoning, see M. Weiss, "Urban Land Developers and the Origins of Zoning Laws: The Case of Berkeley," *Berkeley Planning Journal* 3 (1986): 7–25.

42. Vance, *Geography and Urban Evolution*, emphasizes Oakland's independence. On the port, see Bagwell, *Oakland*; Dykstra, *History of Oakland*. On labor and Oakland politics, see C. Rhomberg, *No There There: Race, Class, and Community in Oakland* (Berkeley, 2004). Oakland's Chamber of Commerce could brag in 1931 that the city was 90 percent open shop. Oakland Chamber of Commerce, Industrial Bureau, *Industrial Facts*, 11. On Knowland, see G. Montgomery and J. Johnson, *One Step from the White House: The Rise and Fall of Senator William F. Knowland* (Berkeley, 1998); E. Cray, *Chief Justice: A Biography of Earl Warren* (New York, 1997). On Kaiser, see M. Foster, *Henry J. Kaiser: Builder in the Modern American West* (Austin, 1989). On the Greater Oakland movement, see R. Self, *American Babylon: Race and the Struggle for Postwar Oakland* (Princeton, 2003).

43. On metropolitan consolidation, see Scott, *San Francisco Bay*, 134; Anon., "The Bay Basin and Greater San Francisco," *Merchants' Association Review* (Dec. 1907); Anon., "Greater San Francisco Edition," *San Francisco Chronicle* (22 Dec. 1907). Scott emphasizes the failure to unify the region and its costs; for a contending view, see R. Lotchin, "The Darwinian City: The Politics of Urbanization in San Francisco between the World Wars," *Pacific Historical Review* 48 (1979): 357–81.

44. For a comprehensive history of local industry, see M. Purcell, *History of Contra Costa County* (Berkeley, 1940), chaps. 24 and 25. Also see J. Whitnah, *The Story of Contra Costa County, California* (Martinez, 1936); G. Emanuels, *California's Contra Costa County* (Fresno, 1986); and promotional pamphlets from 1887, 1903, 1909, and 1915 held in the Bancroft Library, University of California, Berkeley, e.g., Board of Supervisors, *Contra Costa County: Leading County of the West in Manufacturing* (Martinez, c. 1915). Figures on output from Census of Manufacturers, various years. Shipping figures, Calkins and Hoadley, *Economic and Industrial Survey*, 158.

45. Figures from Purcell, *History of Contra Costa*; Emanuels, *California's Contra Costa*. Company towns were segregated by race and ethnicity, as at Tormey/Selby, Valona/Crockett, and Hercules. On Richmond, see J. Whitnah, *A History of Richmond, California* (Richmond, 1944); E. Davis, *Commercial Encyclopedia of the Pacific Southwest* (Berkeley, 1910–15).

46. Oil and electricity fueled Contra Costa's industrialization. The first oil pipeline, from Bakersfield, arrived in 1903. Meanwhile, thanks to the water resources of the Sierra, use of electricity in California manufacturing outran the rest of United States, being six times greater in 1904, three times greater in 1909, and twice as great in 1919. Gordon, *Employment Expansion*, 99. On electrification, see J. Williams, *Electricity and the Making of Modern California* (Akron, 1997).

47. On fish canning, see A. McEvoy, *The Fisherman's Problem: Ecology and Law in the California Fisheries, 1850–1980* (New York, 1986); K. Davis, "Sardine Oil on Troubled Waters" (Ph.D. diss., University of California, Berkeley, 2002).

48. Investors from Issel and Cherny, *San Francisco*, chap. 2. There were few upstart capitalists in Contra Costa's history. On Vallejo versus San Francisco, see R. Lotchin, *Fortress California, 1910–1961: From Warfare to Welfare* (New York, 1992).

49. Quoted by Scott, *San Francisco Bay*, 137, and Issel and Cherny, *San Francisco*, 42. For more such uplifting rhetoric on regional unity in the 1930s, see Issel and Cherny, *San Francisco*, 50.

Chapter Seven

1. The Census Bureau defined a metropolitan district as follows: "In the case of each city having within its own boundaries 200,000 inhabitants or more, there has been delimited what may be termed a 'metropolitan district,' which includes, in addition to the city itself, only those sections of the adjacent territory [the suburban territory within ten miles of the city boundaries] which may be considered as urban in character." U.S. Census, *Twelfth Census of the United States, 1920, Population I* (Washington, D.C., 1920), 62–70.

2. R. Woods, "Pittsburgh: An Interpretation of Its Growth," *Charities and the Commons* 21 (1909): 531.

3. In this chapter "residential suburbs" refer to municipalities that are largely dormitory communities of Pittsburgh and part of the built-up and contiguous urbanized area. "Industrial towns" are the municipalities centered on one or more manufacturing firms within their boundaries or in adjacent municipalities. "Satellite cities" are industrial and commercial municipalities beyond the built-up contiguous urbanized area.

4. P. Kellogg, "The Pittsburgh Survey," *Charities and the Commons* 21 (1909): 519.

5. This chapter focuses on the metropolitan dimensions of this urban industrial growth. For a discussion including the broader regional character of this growth beyond the metropolitan district, see E. K. Muller, "Metropolis and Region: A Framework for Enquiry into Western Pennsylvania," in *City at the Point: Essays on the Social History of Pittsburgh*, ed. S. Hays (Pittsburgh, 1989), 181–211.

6. Pittsburgh annexed the South Side communities in 1872 and Allegheny City in 1907.

7. J. Holmberg, "The Industrializing Community: Pittsburgh, 1850–1880" (Ph.D. diss., University of Pittsburgh, 1981), and U.S. Census, *Seventh Census of the United States, 1850: Manuscript Schedules of Manufactures: Allegheny, Beaver, Butler, Fayette, Washington, and Westmoreland Counties* (National Archives, microfilm).

8. J. Ingham, *Making Iron and Steel: Independent Mills in Pittsburgh, 1820–1920* (Columbus, 1991), 21–46; H. Livesay, *Andrew Carnegie and the Rise of Big Business* (Boston, 1975), 77–90; P. Temin, *Iron and Steel in Nineteenth-Century America* (Cambridge, 1964), 99–121; F. Couvares, *The Remaking of Pittsburgh: Class and Culture in an Industrializing City, 1877–1919* (Albany, 1984), 9–30.

9. J. Bridge, *The Inside History of the Carnegie Steel Company: A Romance of Millions* (Pittsburgh, 1903; 1991), 1–60; J. Wall, *Andrew Carnegie* (Pittsburgh, 1970; 1989), 227–306.

10. When Allegheny City, by far the largest of Pittsburgh's adjacent towns, was annexed by the city in 1907, it became known as the North Side.

11. Ingham, *Making Iron and Steel*, 30–32. While Hazelwood was then a suburb, its contiguity with Pittsburgh and annexation by the city in 1868 made it part of the pre-1870s industrial pattern. J. Tarr and D. DiPasquale, "The Mill Town in the Industrial City: Pittsburgh's Hazelwood," *Urbanism Past and Present* 7 (1982): 2.

12. Holmberg, "Industrializing Community"; J. Tarr, *Transportation Innovation and Changing Spatial Patterns in Pittsburgh, 1850–1934* (Chicago, 1978), 1–11; U.S. Census, *Ninth Census of the United States, 1870, Population I* (Washington, D.C., 1872), 243–56; U.S. Census, *Ninth Census of the United States, 1870: Manuscript Schedules of Manufactures: Westmoreland County* (National Archives, microfilm).

13. Temin, *Iron and Steel*, 125–74; Wall, *Andrew Carnegie*, 307–30.

14. A. D. Chandler, *The Invisible Hand: The Managerial Revolution in American Business* (Cambridge, 1977), 258–69; Ingham, *Making Iron and Steel*, 47–95, 128–56; D. Meyer, "The Rise of the Industrial Metropolis: The Myth and the Reality," *Social Forces* 68 (1990): 731–52; J. Stilgoe, *Metropolitan Corridor: Railroads and the American Scene* (New Haven, 1983), 74–103.

15. The town that grew up around the Westinghouse Electric Manufacturing Company plant was called East Pittsburgh.

16. *Third Industrial Directory of Pennsylvania, 1919* (Harrisburg, 1920), 527, 533; A. Warren, *George Westinghouse, 1846–1914: A Tribute* (n.p., 1914), 27–28.

17. D. Wollman and D. Inman, *Portraits in Steel: An Illustrated History of Jones & Laughlin Steel Corporation* (Kent, 1999), 53–83.

18. M. Magda, *Monessen: Industrial Boomtown and Steel Community, 1898–1980* (Harrisburg, 1985), 3–16.

19. G. D. Smith, *From Monopoly to Competition: The Transformation of Alcoa, 1888–1986* (Cambridge, 1988), 30–32; C. Carr, *ALCOA: An American Enterprise* (New York, 1952), 42–43; G. Fitzsimons and K. Rose, eds., *Westmoreland County, Pennsylvania: An Inventory of Historic Engineering and Industrial Sites* (Washington, D.C., 1994), 221–22; and *Third Industrial Directory*, 1087–88.

20. A. Mosher, "Capital Transformation and the Restructuring of Place: The Creation of a Model Industrial Town" (Ph.D. diss., Pennsylvania State University, 1989), 158–59, and "'Something Better Than the Best': Industrial Restructuring, George McMurtry, and the Creation of the Model Industrial Town of Vandergrift, Pennsylvania, 1883–1901," *Annals of the Association of American Geographers* 85 (1995): 84–107.

21. Wollman and Inman, *Portraits in Steel*, 64–79; D. Brody, *Steelworkers in America: The Nonunion Era* (New York, 1960), 123, 231.

22. P. Krause, *The Battle for Homestead, 1880–1892: Politics, Culture, and Steel* (Pittsburgh, 1992); and M. Byington, *Homestead: The Households of a Mill Town* (New York, 1910), 12–32.

23. T. Mellon, *Thomas Mellon and His Times* (Pittsburgh, 1994).

24. E. Muller, "Westmoreland County Historical Overview," in Fitzsimons and Rose, *Westmoreland County*, 18–19; *Third Industrial Directory*.

25. These data were prepared by researchers at the Historical Society of Western Pennsylvania for a 1998 exhibit on glass manufacturing entitled "Glasstown, U.S.A."

26. R. O'Connor, "The Window Glass Industry and the Expansion of the Pittsburgh Metropolitan Area, 1880–1910" (unpublished paper, Pittsburgh, 1994).

27. K. Warren, *Wealth, Waste, and Alienation: Growth and Decline in the Connellsville Coke Industry* (Pittsburgh, 2001), 1–24.

28. Temin, *Iron and Steel*, 51–98; and Warren, *Wealth, Waste and Alienation*, 25–30.

29. J. Enman, "The Relationship of Coal Mining and Coke Making to the Distribution of Population Agglomerations in the Connellsville (Pennsylvania) Beehive Coke Region" (Ph.D. diss., University of Pittsburgh, 1962); K. Warren, "The Business Career of Henry Clay Frick," *Pittsburgh History* 73 (1990): 3–15; Warren, *Wealth, Waste, and Alienation*, 263; F. Quivik, "The Connellsville Coke Region: A Report by the Historic American Engineering Record with Emphasis on the H. C. Frick Coke Company" (Washington, D.C., unpublished report, 1992).

30. Livesay, *Andrew Carnegie*, 119–82; Wall, *Andrew Carnegie*, 478–536, 714–64; K. Warren, *Triumphant Capitalism: Henry Clay Frick and the Industrial Transformation of America* (Pittsburgh, 1996), 22–51.

31. Warren, *Wealth, Waste, and Alienation*, 263; Quivik, "The Connellsville Coke Region."

32. Warren, *Wealth, Waste, and Alienation*, 195–203; Muller, *Historical Overview*, 5–7, 14–15.

33. The City of Pittsburgh's boundaries for these calculations are consistent with its 1907 boundaries after its last major annexation. (The city added a few small municipalities in the 1920s.) Despite active annexation between the 1840s and early 1900s, Pittsburgh's central city remained a relatively small proportion of the metropolitan area in comparison to cities of similar size across the nation. This fact reflects both minimal annexations after 1907, when suburban communities refused to be absorbed, and the vast extent of the city's industrial suburbs. See G. McLaughlin, *Growth of American Manufacturing Areas: A Comparative Analysis with Special Emphasis on Trends in the Pittsburgh District* (Pittsburgh, 1938), 129.

34. *Ninth Census*, 243–56; *Twelfth Census*, 329–49.

35. *Twelfth Census*, 329–49; U.S. Census, *Fourteenth Census of the United States, 1920, Population*, I (Washington, D.C., 1922), 586–601.

36. E. Muller, "The Pittsburgh Survey and 'Greater Pittsburgh': A Muddled Metropolitan Geography," in *Pittsburgh Surveyed: Social Science and Social Reform in the Early Twentieth Century*, ed. M. Greenwald and M. Anderson (Pittsburgh, 1996), 69–87.

37. P. Kellogg, "The Civic Responsibilities of Democracy in an Industrial District," *Charities and the Commons* 21 (1909): 630.

38. E. Muller, "Industrial Complexes in Metropolitan Formation, 1870–1920" (unpublished paper, University of Pittsburgh, 1994); A. Scott, *Metropolis: From the Division of Labor to Urban Form* (Berkeley, 1988), 33–38, 61–63.

39. Ingham, *Making Iron and Steel*, 140–51.

40. J. Ingham, *Iron Barons: A Social Analysis of an American Urban Elite* (Westport, 1978); Ingham, *Making Iron and Steel*.

41. *Third Industrial Directory, 1919*.

42. D. Hounshell, M. Samber, and J. Tarr, "Technology and Transformation: The Railroad as a Shaper of Regional Space, Natural Resources, Industrial Practice, and Economic Development in Pittsburgh, 1830s–1920" (unpublished paper, Carnegie Mellon University, 1994), 25.

43. Ibid., 47–54.

44. O'Connor, "Window Glass Industry," 28.

45. M. Samber, "Networks of Capital: Creating and Maintaining a Regional Industrial Economy in Pittsburgh, 1865–1919" (Ph.D. diss., Carnegie Mellon University, 1995).

46. W. Schusler, "The Economic Position of Railroad Commuter Service in the Pittsburgh District: Its History, Present, and Future" (Ph.D. diss., University of Pittsburgh, 1958).

47. Quivik, "Connellsvile Coke Region."

48. Brody, *Steelworkers in America*, 50–79.

49. D. Kanitra, "The Westinghouse Strike of 1916" (seminar paper, University of Pittsburgh, 1971).

50. D. Jardini, "In the Footsteps of the Giant: Industrialization of the Lower Connellsville Coal Region" (unpublished paper, Carnegie Mellon University, 1994).

51. R. D. McKenzie, *The Metropolitan Community* (New York, 1933), 71.

52. McKenzie, *Metropolitan Community*, 45.

53. For the definition of yet another kind of suburban community in this period (including one in the Pittsburgh metropolitan area), see J. Borchert, "Residential City Suburbs," *Journal of Urban History* 22 (1996): 283–307.

Chapter Eight

Acknowledgments: I wish to thank Robert Lewis for encouraging me to research and write this chapter and for patiently moving me along. Johanne Sanschagrin and Mark Fram did excellent jobs as research assistants and the latter as cartographer. Grace Chung did a great job putting hasty notes and clumsy tables into order. Leila Gad edited several draft copies and challenged my observations and logic with skill.

1. G. Gad, "Location Patterns of Manufacturing: Toronto in the Early 1880s," *Urban History Review* 22 (1994): 113–38.

2. Current research on industrial location dynamics between the 1850s and the 1890s based on Toronto City Directory data.

3. Exceptions are D. Beeby, "Industrial Strategy and Manufacturing Growth in Toronto, 1880–1910," *Ontario History* 76 (1984): 199–232; G. D. Garland, "Suburbanization and Transition to Monopoly Capitalism" (master's thesis, University of Toronto, 1978); R. Harris, *Unplanned Suburbs: Toronto's American Tragedy, 1900–1950* (Baltimore, 1996), esp. 51–85.

4. "Suburbanization" and "decentralization" are predicated on an earlier "centralization" of manufacturing. However, to conceptualize manufacturing location as having been largely or exclusively central at an earlier point in time may not be empirically sustainable.

5. For a discussion of single-story and multistory factory design, see R. Lewis, "Redesigning the Work Place: The North American Factory in the Inter War Period," *Technology and Culture* 42 (2001): 665–84.

6. A. Scott, "Locational Patterns and Dynamics of Industrial Activity in the Modern Metropolis," *Urban Studies* 19 (1982): 111–42, and *Metropolis: From the Division of Labor to Urban Form* (Berkeley, 1988).

7. Some discussions of the links between place of residence and place of work can be found in Scott, *Metropolis*, chap. 7, on local labor markets.

8. Garland, "Suburbanization and Transition," based on emerging Marxian conceptions of suburbanization such as D. Gordon, "Capitalist Development and the History of American Cities," P. Ashton, "The Political Economy of Suburban Development," and A. Markusen, "Class and Urban Social Expenditure: A Marxist Theory of Metropolitan Government," all in *Marxism and the Metropolis: New Perspectives in Urban Political Economy*, ed. W. Tabb and L. Sawers (New York, 1978), 25–63, 64–89, and 90–111, respectively.

9. J. Lemon, *The Toronto Habour Plan of 1912: Manufacturing Goals and Economic Realities* (Toronto, 1990), casts doubt on the importance of water transport.

10. The discussion of Toronto's industrial structure over a seventy-year time span is very crude. Industrial classifications changed frequently and in Table 8.1 all earlier classes were aggregated to the industrial classification used by the Dominion Bureau of Statistics in 1951. Industrial sectors shown in Table 8.1 often consisted of very diverse industries, and the character of many sectors changed over time. For instance, the transportation equipment sector included carriage building in 1880 and carriage building and automobile manufacturing in 1911; by 1951 this sector consisted largely of the automobile industry.

11. J.M.S. Careless, *Toronto to 1918: An Illustrated History* (Toronto, 1984), 157–58; J. Lemon, "Toronto among North American Cities," in *Forging a Consensus: Historical Essays on Toronto*, ed. V. Russell (Toronto, 1984), 323–51, esp. 330.

12. Harris, *Unplanned Suburbs*, esp. 200–232; B. Myrvold, *The People of Scarborough: A History* (Scarborough, Ontario, 1997), 115–16, 136–37.

13. Employment in 1880–81 was 12,708, according to the *Census of Canada, 1880–81*, vol. 3, 323–496; employment in 1915 was 72,798, according to *Postal Census of Manufactures, Canada, 1916* (Ottawa, 1917), 190.

14. On the National Policy of 1879 and its implications, see M. Bliss, *Northern Enterprise: Five Centuries of Canadian Business* (Toronto, 1987), 285–43; I. Wallace, *A Geography of the Canadian Economy* (Toronto, 2000), 10–11, 19–20, 72–73.

15. E. Bloomfield, G. Bloomfield, and M. Valliers, "Urban Industrial Development in Central Canada: Manufacturing Structure, 1911," in *Historical Atlas of Canada*, vol. 3, ed. D. Kerr and D. Holdsworth (Toronto, 1990), plate 13.

16. E. Neufeld, *A Global Corporation: A History of the International Development of Massey-Ferguson Limited* (Toronto, 1969). The Massey Manufacturing Company became Massey-Harris in 1894 and Massey-Ferguson in 1953.

17. G. Bloomfield et al., "The Changing Structure of Manufacturing," in Kerr and Holdsworth, *Historical Atlas*, plate 7.

18. R. Stamp, *Bright Lights, Big City: The History of Electricity in Toronto* (Toronto, 1991). Evidence of individual factory-generated electricity appears in various company histories.

19. The Northey Pump Company is usually credited with having erected Toronto's first single-story factory in 1892. See Beeby, "Industrial Strategy," 215; Harris, *Unplanned Suburbs*, 56. However, the Inglis machinery company probably built a single-story factory in 1885, and there may have been even earlier ones in the heavy engineering and metals industries. On Inglis see D. Sobel and S. Meurer, *Working at Inglis: The Life and Death of a Canadian Factory* (Toronto, 1994), 18.

20. G. Ferguson, "Decentralization of Industry and Metropolitan Control," part 2, *Journal of the Town Planning Institute of Canada* 2 (1923): 8.

21. M. Doucet, "Mass Transit and the Failure of Private Ownership: The Case of Toronto," *Urban History Review* 3 (1977): 3–33.

22. The 1880s and 1890s seem to have been the decades with the fiercest competition for manufacturing establishments between municipalities in Southern Ontario. At least the literature is on this time period; see Beeby, "Industrial Strategy."

23. M. Laycock and B. Myrvold, *Parkdale in Pictures: Its Development to 1889* (Toronto, 1991), 52–55.

24. My account of West Toronto Junction's history as an industrial suburb relies heavily on Garland, "Suburbanization and Transition," and also on A. de Fort-Menares, *West Toronto Junction Engine House* (Ottawa, 1997), and A. Rice, *West Toronto Junction Revisited*, 4th ed. (Toronto, 1999).

25. The 1906 Census shows (for 1905) 1,842 employees in Toronto Junction in establishments of five and more persons and 1,860 in establishments irrespective of size. Canada, Census and Statistics, *Bulletin II, Manufactures of Canada*, 64.

26. M. Denison, *C.C.M.: The Story of the First Fifty Years* (Toronto, 1946).

27. J. Willis, *This Packing Business* (Toronto, n.d. [c. 1963]), 35.

28. A list of manufacturing establishments was derived from the *Toronto City Directory for 1909*. All firms and households are listed under West Toronto in the Suburban Section, 67–97.

29. *Toronto City Directory for 1909*, 67–97; *Insurance Plan of Toronto Junction, 1912*, Sheet 3, with entry of "400 hands."

30. Harris, *Unplanned Suburbs*, 70.

31. Ibid.

32. Ibid.

33. Garland, "Suburbanization and Transition," 134–35.

34. Rice, *West Toronto*, 74.

35. Ibid.

36. Lemon, *Toronto Harbour Plan*; S. Smith, "The Development of Toronto's Crown Reserves as Industrial Areas, 1793–1900" (master's thesis, University of Toronto, 1999).

37. According to the *Census of Manufactures of Canada* taken in 1916 there were 72,798 manufacturing jobs in the city of Toronto in 1915. In the town of Weston, in 1915 a suburb of Toronto, there were 439. Employment numbers for other, smaller suburbs and unincorporated areas are not available. A crude estimate would be in the range of 1,500 to 2,500 manufacturing jobs in the suburban zone, including the 439 in Weston. The suburban share of total urban area manufacturing employment would have only been 2 to 3.3 percent. See Canada, *Postal Census*, 188–91.

38. For 1915 manufacturing employment, see note 37. The 1951 employment is from Canada, *General Review of the Manufacturing Industries of Canada 1951* (Ottawa, 1954), 120. Population numbers are from Harris, *Unplanned Suburbs*, 42. Greater Toronto is the urban area consisting of the city of Toronto and the suburbs as defined in the 1951 Census of Canada as the Toronto Census Metropolitan Area (Figure 8.4). The 1951 Toronto CMA coincided with the Municipality of Metropolitan Toronto created in 1953.

39. From 1917 on the government of Canada's statistical agency, the Dominion Bureau of Statistics, conducted annual surveys of manufacturers. The results were published in the series *Manufacturing Industries of Canada* and in the *Canada Year Books*. This series included data for cities and towns above certain population thresholds or

aggregate-value-of-production thresholds. As suburban municipalities reached these thresholds, they appeared in the published record. Data for rural townships were never published and the urban area totals were not available either. However, from 1944 on the *Manufacturing Industries of Canada* included data for Greater Toronto as a whole; the establishment and employment numbers for the rural townships can be calculated as a residual (Table 8.4). For the definition of Greater Toronto, see note 38.

40. Harris, *Unplanned Suburbs*, 200–232; E. Heyes, *Etobicoke: From Furrow to Borough* (Etobicoke, Ontario, 1974), 128; Myrvold, *Parkdale*, 113–15.

41. F. Frisken, "A Triumph for Public Ownership: The Toronto Transit Commission 1921–1953," in Russell, *Forging a Consensus*, 238–71.

42. Ibid., 239, 251.

43. E. Bloomfield, "Municipal Bonusing of Industry: The Legislative Framework in Ontario to 1930," *Urban History Review* 9 (1981): 59–76.

44. Harris, *Unplanned Suburbs*, 45–50.

45. Dension, *C.C.M*, 35.

46. The history of the Massey-Harris Weston branch plant has been pieced together from M. Denison, *Harvest Triumphant: The Story of Massey-Harris* (Toronto, 1949); Neufeld, *Global Corporation*; and F. Hotson, *DeHavilland in Canada* (Toronto, 1999).

47. G. Adam, *Toronto Old and New* (Toronto, 1891), 183; Beeby, "Industrial Strategy," 204.

48. Adam, *Toronto*, 185; Beeby, "Industrial Strategy," 204.

49. Based on systematic scrutiny of the New Toronto and Mimico sections of the Toronto City Directories, 1895 to 1915.

50. Harris, *Unplanned Suburbs*, 59–63.

51. D. Campbell, *Global Mission: The Story of Alcan*, (n.p., 1985), 1:367–69; A. Whitaker, *Aluminium Trail* (Montreal, 1975), 250–52.

52. Harris, *Unplanned Suburbs*, 79–82.

53. Heyes, *Etobicoke*, 145.

54. My account of the history of manufacturing in Leaside in the context of urban development relies on L. McCann, "Planning and Building the Corporate Suburb of Mount Royal, 1910–1925," *Planning Perspectives* 11 (1996): 259–301; Leaside Town Council, *The Story of Leaside* (Leaside, Ontario, 1931); J. Pitfield, *Leaside* (Toronto, 1999); and J. Pryce, *The Red Reel: The Story of Canada Wire* (Toronto, 1978); K. Rennick, "The Town of Leaside: Origins and Development, 1913–1939" (Geography Research Project, University of Toronto, 1986); J. Rempel, *The Town of Leaside: A Brief History* (Toronto, 1982).

55. Leaside Town Council, *Story of Leaside*, 19.

56. A systematic account of government investment in wartime production facilities, both government and privately operated, is provided in Canada, Department of Munitions and Supply, *Government Financed Expansion of Industrial Capacity in Canada as at December 31, 1943* (Ottawa, 1944). The report lists companies but not locations.

57. Harris, *Unplanned Suburbs*, 58–59.

58. P. Schulz, *The East York Workers' Association: A Response to the Great Depression* (Toronto, 1975).

59. Information on DeHavilland in Toronto is from Hotson, *DeHavilland*.

60. R. Bonis, *A History of Scarborough* (Scarborough, Ontario, 1968), 206–9; L. Hurst, "Lipsticks and Bombshells," *Toronto Star* (31 Aug. 1989), J10; R. Schofield, "Crumbling Tunnel Reminder of War Effort," *Toronto Star* (7 Feb. 1991), E5.

61. Ajax Historical Board, *The Pictorial History of Ajax* (Ajax, Ontario, 1941; 1972).

62. World War II industrial production in Toronto Township is covered briefly in R. Riendeau, *Mississauga: An Illustrated History* (n.p., 1985). A more detailed account of Victory Aircraft Limited (and its predecessor, National Steelcar Company, and successor, A. V. Roe Canada) can be pieced together from G. Stewart, *Shutting Down the National Dream: A. V. Roe and the Tragedy of the Avro Arrow* (Toronto, 1988).

63. Neufeld, *Global Corporation*, 53–59.

64. Hotson, *DeHavilland*, 101.

65. D. Hiebert, "Discontinuity and the Emergence of Flexible Production: Garment Production in Toronto, 1901–1931," *Economic Geography* 66 (1990): 228–53.

66. Harris, *Unplanned Suburbs*, 65–85; R. Harris and A. Bloomfield, "The Impact of Industrial Decentralization on the Gendered Journey to Work, 1900–1940," *Economic Geography* 73 (1997): 94–117.

67. B. Donald, "Spinning Toronto's Golden Age: The Making of a 'City That Worked,'" *Environment and Planning A* 34 (2002): 2132–36.

Chapter Nine

Acknowledgments: The author thanks Todd Gish and Kathy Kolnick for research assistance, the Southern California Studies Center (USC) for research support, and Robert Lewis, David Meyer, Ted Muller, and Dick Walker for incisive comments on drafts.

1. "The Undiscovered City—To All Wonders of Southern California, Add Another: Industrial Los Angeles Turns Out More Dollar Product than Pittsburgh," *Fortune* 39 (1949): 76–83, 148, 150, 153–54, 156, 158, 160; quotations at 78, 82.

2. The magnitude and nature of change in California during the war years is a topic of scholarly debate. For an introduction, see R. Lotchin, "World War II and Urban California: City Planning and the Transformation Hypothesis," *Pacific Historical Review* 62 (1993): 143–71, a critique of G. D. Nash, *The American West Transformed: The Impact of the Second World War* (Bloomington, 1985), and M. S. Johnson, *The Second Gold Rush: Oakland and the East Bay in World War II* (Berkeley, 1993). See also Lotchin, *Fortress California, 1910–1961: From Warfare to Welfare* (New York, 1992).

3. In "Symbolic Landscapes: Some Idealizations of American Communities," Donald Meinig fixed southern California as the culture hearth for the third of three symbolic American landscapes: single-family housing on broad, landscaped lots sited along curved streets with direct access to freeways. Meinig overstates the importance of a landscape type that accounts for a small percentage of regional development. Meinig, ed., *The Interpretation of Ordinary Landscapes* (New York, 1979), 169. For overviews of the literature, see M. Marsh, "Reconsidering the Suburbs: An Exploration of Suburban Historiography," *Pennsylvania Magazine of History and Biography* 112 (1988): 580–605; and J. Phalen, *The Suburbs* (New York, 1995). See also the introduction in G. Hise, *Magnetic Los Angeles: Planning the Postwar Metropolis* (Baltimore, 1997).

4. Hise, *Magnetic Los Angeles*, 82.

5. See companion essays in this volume. Also see *American Quarterly* 46 (1994) and the exchange between K. Jackson, R. Fishman, and R. Harris in *Journal of Urban History* 13 (1987) and 15 (1988). For Los Angeles, see Hise, *Magnetic Los Angeles*; R. Longstreth, *From City Center to Regional Mall: Architecture, the Automobile, and Retailing in Los Angeles, 1920–1950* (Cambridge, 1997); B. Nicolaides, *My Blue Heaven: Life and Politics in the Working-Class Suburbs of Los Angeles, 1920–1965* (Chicago, 2002); N. Quam-Wickam, "'Petroleocrats and Proletarians': Work, Class, and Politics in the Southern California Oil Industry, 1917–1935" (Ph.D. diss., University of California, Berkeley, 1993).

6. The Chamber of Commerce used this logotype from at least 1914. See its *Manufacturers Directory and Commodity Index* from that year and subsequent editions. The Chamber's industrial department adopted the slogan for books and pamphlets; see, for example, *Los Angeles—Nature's Workshop, The Home of Efficient Labor* (Los Angeles, 1921), and *Dynamic Los Angeles County* (Los Angeles, 1938).

7. C. McWilliams, *Southern California: An Island on the Land* (Salt Lake City, 1946; 1990), chaps. 12 and 16; R. Fogelson, *The Fragmented Metropolis: Los Angeles, 1850–1930* (Cambridge, 1967), esp. chap. 6; K. Starr, *Material Dreams: Southern California through the 1920's* (New York, 1990); M. Davis, *City of Quartz: Excavating the Future in Los Angeles* (London, 1990), 118–20; R. Romo, *History of a Barrio: East Los Angeles* (Austin, 1983), chap. 4.

8. For an early critique of this ideology, see W. Woehlke, "Smoke-stacks on the Pacific," *Sunset* 31 (1931): 1161–71. For a contemporary assessment that introduces race as a central factor in labor hiring practices, see C. Johnson, "Industrial Survey of the Negro Population of Los Angeles, California" (Department of Research and Investigations of the National Urban League, 1926). In *The Productivity of Labor in the Rubber Tire Manufacturing Industry* (New York, 1940), J. Gaffey asserts that labor savings were not a critical factor for establishing West Coast operations and the selection of southern California for branch plants.

9. The Chamber published a running account of firms establishing plants. See the annual reports produced by the industrial department as well as articles in *Southern California Business*, such as "Bringing New Industries to Town" (1925), 16–17. According to the California State Mining Bureau, approximately 10 million pounds of lead valued at roughly $700,000 were mined in the state during 1923, *California Mineral Production for 1923*, cited in W. French, "A Study of Locational Factors for Industrial Plants in and about Los Angeles, California," (MBA thesis, University of Southern California, 1926), 95.

10. "City Industrial Tract," report for A. Arnoll by H. Lafler, Sales Agent, Walter H. Leimert Co, (c. 1923), in Los Angeles Area Chamber of Commerce collection, Department of Special Collections, University of Southern California, box 74 (hereafter LAACC Collection).

11. French, "Study of Locational Factors" chap. 4; quotation at 23.

12. Eberle & Riggleman Economic Service, "The Industrial Land Situation in Los Angeles," *Weekly Letter* (28 Dec. 1925), and "Some Considerations of the Los Angeles Industrial Situation," *Weekly Letter* (21 Sept. 1925, no pagination). This concern was a constant. Fifteen years later planner Bryant Hall noted a "disturbing" ratio of industrial land relative to population in the county. See the tabulations and annotations in a joint WPA/Regional Planning Commission *Land Use Analysis* (1940), part

of a WPA-sponsored survey of land use and population. Binder in the Regional Planning Collection, The Huntington Library, San Marino, California (hereafter The Huntington).

13. City Industrial Tract, 4.

14. M. Davis, "Sunshine and the Open Shop: Ford and Darwin in 1920s Los Angeles," in *Metropolis in the Making: Los Angeles in the 1920s*, ed. T. Sitton and W. Deverell (Berkeley, 2001).

15. *Los Angeles Times* editorial, special section, "Midwinter Edition" (1 Jan. 1905).

16. Los Angeles City Directory Co., *Los Angeles City Directory, 1905*.

17. See the material in the California Ephemera Collection, Department of Special Collections, UCLA, box 39, folder "Industry-Historical-CA-1" (hereafter UCLA Ephemera). Quotation from brochure, "The Development of an Idea." WPA researchers collected these documents for "The Industrial Section" in Workers of the Writers' Program of the Work Projects Administration, *Los Angeles: A Guide to the City and Its Environs* (New York, 1941; 1951), 167. See also A. Bynon and Co., *Los Angeles City and County Directory, 1886–1887* (Los Angeles, 1887), 164, and K. Doyle, "A Factory Here since 1860," *Southern California Business* 10 (1931): 14; "Soap Company Plans New Five-story Building," *Los Angeles Times* (3 Sept. 1922), sec. 5, 4.

18. "The Wholesale Terminal—Efficiency For: The Wholesaler, Jobber, Shipper, Merchant, Manufacturer, Consumer" (c. 1916); "Los Angeles—A Good Place to Manufacturer or Warehouse" (n.d.); "Los Angeles Union Terminal Company, Terminal Refrigeration Company, Union Terminal Warehouse" (n.d.), all in UCLA Ephemera, box 39, folder "Industry-Historical-CA-1." "Bringing the Railway and Its Freight Customer Together," *Electric Railway Journal* 50 (7 July 1917): 6–7.

19. Oxnam quotation in "Los Angeles Tenement Problem," Municipal League *Bulletin* 3 (1925): 5.

20. French, "Study of Locational Factors." See also the tract maps in The Huntington Library ephemera collection, Eph J3–4(5), as well as the land-use maps the WPA produced in collaboration with the Los Angeles Regional Planning Commission (vol. 6, sheet 37), also at The Huntington.

21. CCIH, "A Community Survey Made in Los Angeles City," (San Francisco, 1919); "Industrial and Business Property," *Los Angeles Times* (19 Jan. 1926), sec. 2, 17. In her "Aircrafters" column, published in *The War Worker* 1 (2d half, July 1943): 2, Esther Beverly Owens noted that the Lockheed Aircraft Plant 7, located on Seventh Street "across the bridge from Santa Fe [Avenue]," was "thickly populated with American Negroes, Mexicans, and members of other groups" busy wiring pivot casings and drilling subassemblies. In "Industrial Strength," *Los Angeles Times* (3 Aug. 1997), D1, the author takes readers through a "200-block area east of L.A.'s skyscrapers" with aging buildings, razor-wire, and homeless encampments, "fertile ground for foreign trade and immigrant entrepreneurs" engaged in the garment, printing, produce, and wholesale sectors.

22. For a discussion of pollution in turn-of-the-century Los Angeles, see D. Johnson, "A Serpent in the Garden: Institutions, Ideology, and Class in Los Angeles Politics, 1901–1911" (Ph.D. diss., UCLA, 1996), esp. chap. 3.

23. The history of zoning in Los Angeles, and the effect of these regulations for zoning in other cities and national guidelines, has yet to be written. For a contemporary assessment of the significance of these statutes, see L. Veiller, "City Planning in

Los Angeles," *The Survey* 26 (22 July 1911): 599–600. For the district boundaries, see Ordinance N.17135 (new ser., 1908), Ordinance N.17136 (new ser., 1908), and Ordinance N.19500 (new ser., 1909).

24. "Industrial Lands," in the LAACC *Members' Annual* (1929): 82. Agenda item (14 Dec. 1922), 6, in *Stenographer's Notes, 1922*. Both items in the LAACC Collection, box 18.

25. On labor-management, strikes, and the 1910 bombings, see G. Stimson, *Rise of the Labor Movement in Los Angeles* (Berkeley, 1955), esp. chaps. 19 and 21.

26. A. G. Arnoll, "Balanced Prosperity in the Los Angeles Area," a presentation sponsored by the Inter-City and Suburbanization Committee of the Los Angeles Municipal League. Printed in the Municipal League *Bulletin* 6 (March 1924). At the time, Arnoll was assistant secretary for the LAACC and director of its industrial department.

27. "Torrance, The Model Industrial City," UCLA Ephemera, box 103, folder "Torrance."

28. "Manufacturing Committee—Westinghouse Track Connections," in Los Angeles Area Chamber of Commerce, *Stenographer's Notes, Board of Directors Meetings, 1922*, 2–5, LAACC Collection, box 18.

29. G. Law quoted in P. Sheehan, *Hollywood as a World Center* (Hollywood, 1924), 20–21.

30. See Fogelson, *Fragmented Metropolis*, esp. 123–29; "Balanced Progress," *Los Angeles Times* (18 Nov. 1923), sec. 2, 4.

31. "Balanced Progress," *Los Angeles Times*.

32. H. Allen, *The House of Goodyear* (Akron, 1936); Goodyear Tire and Rubber Company of California, *Three Dynamic Decades in the Golden State, 1920–1950* (c. 1950), in Los Angeles Examiner Collection, Regional History Center, USC, photo folder "Goodyear"; P. Rhode, "California's Emergence as the Second Industrial Belt: The Pacific Coast Tire and Automobile Industries" (unpublished paper, University of North Carolina, 1994); "Civic Heads, U.P. Officials Aid Ceremony, Los Angeles May Outstrip Akron Soon," *Los Angeles Examiner* (24 Jan. 1929), 1; "Tire Manufacture a Major Industry Here," *Industrial Los Angeles County* 2 (May 1930): 4; "Rubber Industry—Los Angeles County," editorial in *Industrial Los Angeles County* 2 (May 1930).

33. F. Kidner and P. Neff, *An Economic Survey of the Los Angeles Area* (Los Angeles, 1945).

34. Here I am applying an interpretation drawn from D. Holdsworth's assessment of Chicago in "The Invisible Skyline," *Antipode* 26 (1994): 141–46.

35. Plates 14–17 bound as an appendix to *Report and Recommendations on a Comprehensive Rapid Transit Plan for the City and County of Los Angeles* (Chicago, 1925).

36. William Daum Jr., interview by author, Daum Company office, 123 South Figueroa, Los Angeles, 19 June 1996; and clippings in office scrapbooks.

37. See materials in the folder "Los Angeles Streets" in the Henry Z. Osborne Papers, Department of Special Collections, USC. Quotation from "Program Is Outlined, East Side Organization Names Committees and Prepares to Campaign for Improvements," *Los Angeles Times* (18 Nov. 1929), sec. 5.

38. "The Cat's Out of the Bag," a six-column advertisement in the *Los Angeles Times* (10 Sept. 1922), sec. 5, 3. For Belvedere Gardens, see "Ready to Open New Belvedere Gardens Tract," *Los Angeles Times* (10 April 1921), sec. 5, 3.

39. Central Manufacturing District, Inc., "Central Manufacturing District of Los Angeles: A Book of Descriptive Text, Photographs and Testimonial Letters about the Central Manufacturing District of Los Angeles—'The Great Western Market,'" (Aug. 1923), LAACC Collection; "Great Industrial City Here Being Created by the Central Manufacturing District," *Los Angeles Times* (15 July 1923); "Cabbage Patch to Industrial Paradise," *Santa Fe Magazine* (Nov. 1929), 21–25; H. Poronto, "How Chicago Came to Los Angeles Told by Head of Central Manufacturing Dist.," *Southwest Builder and Contractor* 62 (13 July 1923): 34.

40. On the sale, see "Los Angeles Holdings Go to Rich Group," *Los Angeles Examiner*, 7 March 1928, and "Santa Fe Buys Industrial Hub for $15,000,000," *Los Angeles Examiner*, 11 April 1929. For city biographies of Bell and other incorporated communities, see Los Angeles Chamber of Commerce, Industrial Department, *Industrial Communities of Los Angeles Metropolitan Area* (Los Angeles, 1925), and *California Real Estate Magazine* 10 (June 1930), a special issue devoted to "Growth and Progress of the Golden West."

41. On the plague, see W. Deverell, "Plague in Los Angeles, 1924: Ethnicity and Typicality," in *Over the Edge: Remapping Western Experiences*, ed. V. Matsumoto and B. Allmendinger (Berkeley, 1998), and a photographic collection at the Bancroft Library, University of California, Berkeley, available on the Californian Heritage site at http://sunsite.berkeley.edu/Collections/. This was an ongoing process. For a post–World War II account, see "Farewell to 'Mañana': Hicks Camp Prepares to Abandon Old Ways," *Los Angeles Times* (13 May 1949), sec. 3, 6, which describes how an enclave of approximately 150 farmworker families living in "El Monte's bit of old Mexico" were being forced out by property owner Harvey Youngblood, who planned to develop the site as an industrial park for light manufacturing.

42. Los Angeles Board of City Planning Commissioners, "Subdividing of Land," in *Annual Report, 1929–1930* (Los Angeles, 1930), 49.

43. WPA/Regional Planning Commission, *Land Use Analysis: Final Report* (Los Angeles, 1941).

44. County of Los Angeles, "Proceedings of the First Regional Planning Conference of Los Angeles County" (Los Angeles, 1922), 6.

45. F. Viehe, "Black Gold Suburbs: The Influence of the Extractive Industry on the Suburbanization of Los Angeles, 1890–1930," *Journal of Urban History* 8 (1981): 3–26.

46. For firm locations, see the LAACC Industrial Bureau Publications, *Manufacturers' Directory and Commodity Index*. I consulted the second (1915) and fifth (1920) editions. On Harry H. Culver, see a biography ("as of Sept. 1, 1929") and the clippings in the Examiner Collection, Department of Special Collections, USC. For a secondary account of the film industry, see S. Christopherson and M. Storper, "The City as Studio: The World as Back Lot: The Impact of Vertical Disintegration on the Location of the Motion Picture Industry," *Environment and Planning D: Society and Space* 4 (1986): 305–20.

47. Department of Public Safety, *Second Annual Police Benefit Book* (Culver City, 1927).

48. This section is drawn from Hise, *Magnetic Los Angeles*, chap. 4.

49. D. Hansen (McDonnell Douglas, Long Beach), letter to author (3 June 1994). See also F. Cunningham, *Skymaster: The Story of Donald Douglas* (Philadelphia, 1943),

and C. Maynard, *Flight Plan for Tomorrow: The Douglas Story, A Condensed History* (Santa Monica, 1962).

50. "Finest Community Development in 20 Years" and "Typical Homes in Westchester District," *Los Angeles Evening Herald and Express* (28 March 1942). See also "City Planners Flock to Study Westchester," *Los Angeles Daily News* (8 May 1942), 27, and the low-altitude oblique aerials of this development in the Spence and Fairchild Aerial Photo Collections in the Department of Geography, UCLA.

51. R. Williams, *The Country and the City* (New York, 1980); W. Cronon, "The Trouble with Wilderness; Or, Getting Back to the Wrong Nature," in *Uncommon Ground: Rethinking the Human Place in Nature*, ed. W. Cronon (New York, 1996); CCIH, *A Community Survey Made in Los Angeles* (San Francisco, 1919), 23.

Chapter Ten

1. J. Arnold, *The New Deal in the Suburbs* (Columbus, 1971), 11–12.

2. The role of technology in furthering urbanization was emphasized by historian Lewis Mumford, who referred to this spread of settlement as the "Fourth Migration," following the westward movement of pioneers, the rise of industrialization, and the growth of centralized cities. Lewis Mumford, "The Fourth Migration," *Survey Graphic* 54, special issue (1 May 1925).

3. R. D. McKenzie, *The Metropolitan Community* (New York, 1933), 71.

4. "Following upon the adoption of the automobile as a distance-reducer, the cityman's appreciation of the desirability of the invigorating fresh country air took a decided up-turn. The realtor capitalized the situation, and our national magazines contributed their effective influence to the new enthusiasm for the soil, the flowers and all of the out-of doors." E. Van Cleef, "The World's Greatest Migration," *American City* 39 (Sept. 1928): 154.

5. U.S. Bureau of the Census, *Population: 1930* (Washington, D.C., 1931), 1131; A. Nevins and F. Hill, *Ford: Expansion and Challenge* (New York, 1957), 687; P. Reid, *Industrial Decentralization: Detroit Region, 1940–1950 (Projection to 1970)* (Detroit, 1951), 3–4. Here I define the Detroit metropolis by the boundaries set by the 1930 Census, which included the following incorporated places found in parts of Wayne, Macomb, and Oakland Counties: Detroit, Allen Park, Berkley, Birmingham, Bloomfield Hills, Centerline, Clawson, Dearborn, East Detroit, Ecorse, Ferndale, Fraser, Garden City, Grosse Pointe, Grosse Point Farms, Grosse Point Park, Grosse Point Shores, Hamtramck, Highland Park, Huntington Woods, Inkster, Lake Angelus, Lincoln Park, Lochmoor, Melvindale, Mount Clemens, Northville, Oak Park, Pleasant Ridge, Plymouth, Pontiac, River Rouge, Riverview, Rochester, Roseville, Royal Oak, St. Clair Shores, Sylvan Lake, Trenton, Utica, Warren, Wayne, Wyandotte. U.S. Bureau of the Census, *Metropolitan Districts: Population and Area*, ed. Clarence Batschelet (Washington, D.C.: U.S. Government Printing Office, 1932), 72–74.

6. T. Ticknor, "Motor City: The Impact of the Automobile Industry upon Detroit, 1900–1975" (Ph.D. diss., University of Michigan, 1978), 201–4. Ticknor bases this and the following statistics on census data for population and manufacturing. In 1940, 36 percent of Detroit's manufacturing was in its suburbs, and 23 percent of its residential population was suburban.

7. *Detroit City Directory, 1924–25.*

8. *Detroit City Directory, 1925–26;* M. Holli, ed., *Detroit* (New York, 1976), 116–22; O. Zunz, *The Changing Face of Inequality: Urbanization, Industrial Development, and Immigrants in Detroit, 1880–1920* (Chicago, 1982). The population of the city was 285,704 in 1900, 993,675 in 1920, and 1,623,452 in 1940.

9. U.S. Bureau of the Census, *Population: The Growth of Metropolitan Districts, 1900–1940,* ed. Warren Thompson (Washington, D.C., 1947), and *Fifteenth Census of the U.S., Metropolitan Districts: Population and Area* (Washington, D.C., 1932). The population of the metropolis was 1,252,909 in 1920, 2,104,764 in 1930, and 2,295,867 in 1940. Ticknor, "Motor City," 185–205.

10. Zunz, *Changing Face,* 308–9.

11. Some automotive jobs were lost from Detroit to other states as well as to the city's suburbs. Ticknor, "Motor City," 287–88.

12. Nevins and Hill, *Ford,* 202, 201.

13. In describing Dearborn prior to the consolidation of 1928, this study includes Springwells or Fordson, depending on what name it was going by at the time.

14. J. Moore, "History of the Dearborn Area" (unpublished paper, 1957), Bentley Collection, University of Michigan.

15. *Dearborn News* (9 June 1928).

16. A. M. Wibel, "Reminiscences," 1–7, Henry Ford Museum.

17. G. Watson Organization, Inc., "Dearborn Facts and Figures" (pamphlet, 1927), 1, City of Dearborn, History File (hereafter HF), Dearborn Historical Museum (hereafter DHM); Moore, "History of Dearborn," 243.

18. C. Ford, "Henry Ford's Home Town," *Dearborn Magazine* 1 (1926): 2, City of Dearborn, HF, DHM.

19. "The Inside Story of Henry Ford," *Dearborn Magazine* 1 (1926): 3; "Greater Dearborn: The Consolidation Story" (unpublished paper, n.d.), 1–9, City of Dearborn, Consolidation, HF, DHM; S. Shiefman, "Ever Hear of Dearson?" (unpublished paper, 1954), 1, City of Dearborn, Consolidation, HF, DHM.

20. Moore, "History of Dearborn," 299–306; Fordson Board of Commerce, "Fordson, Michigan: Western Gateway to Detroit" (report, 1927), 2–11, City of Fordson, HF, DHM.

21. J. Karmann, "Oral History," 4, DHM.

22. Fordson Board of Commerce, "Fordson," 2–11.

23. I. Becker, "Oral History," 150, DHM.

24. A. Ammerman, "A Sociological Survey of Dearborn" (master's thesis, University of Michigan, 1940), 57–58, DHM.

25. Becker, "Oral History," 149, 74–75.

26. Karmann, "Oral History" 4, 13, 17, 18; *Dearborn News* (9 June 1928).

27. Moore "History of Dearborn," 245. "Greater Dearborn: The Consolidation Story," 1–9; editorial by the Save Dearborn Association, *Dearborn Press* (17 May 1928).

28. Consolidated Cities Association, pamphlet, City of Dearborn, Consolidation, HF, DHM.

29. R. Crabb, *Birth of a Giant* (Philadelphia, 1969), 411; Nevins and Hill, *Ford,* 687; Zunz, *Changing Face,* 293.

30. Ticknor, "Motor City," 110–13.

31. Ticknor, "Motor City," 111; U.S. Bureau of the Census, *Population: 1930,* 1131–32; U.S. Bureau of the Census, *Manufactures: 1929* (Washington, D.C., 1930),

250–51; U.S. Bureau of the Census, *Population: 1940* (Washington, D.C., 1941), 873, 916–17; U.S. Bureau of the Census, *Manufactures: 1939* (Washington, D.C., 1942), 479.

32. U.S. Bureau of the Census, *Population: 1920* (Washington, D.C., 1921), 488–95; Bureau of the Census, *Population: 1930*, 1131–48; Bureau of the Census, *Population: 1940*, 868–73, 916–17.

33. R. Duffus, "Detroit: Utopia on Wheels" *Harper's* 162 (1930), 50–51.

34. Ticknor, "Motor City," 193–97.

35. Ibid., 190–93.

36. R. Thomas, *Life for Us Is What We Make It: Building Black Community in Detroit, 1915–1945* (Bloomington, 1992), 90–92; Ticknor, "Motor City," 237–39.

37. Zunz, *Changing Face*, 327–28.

38. K. B. Eckert, *Buildings of Michigan* (New York, 1993), 28–63.

39. Ticknor, "Motor City," 185–205.

40. C. A. Player, "Detroit: Essence of America," *New Republic* (3 Aug. 1927), 275.

41. J. Welliver, "Detroit, the Motor-car Metropolis," *Munsey's* (1919), 655.

42. "FOB Detroit," *Outlook* (22 Dec. 1915), 979.

43. Ticknor, "Motor City," 185–87.

44. Eckert, *Buildings of Michigan*, 28–63.

45. Federal Writers' Program, Works Progress Administration (WPA), *Michigan: A Guide to the Wolverine State* (New York, 1941), 232, 234.

46. Ticknor, "Motor City," 185–87.

47. R. Fogelson, *Downtown, Its Rise and Fall, 1880–1950* (New Haven, 2001), 224.

48. A. Pound, *Detroit, Dynamic City* (New York, 1940), 359–61.

49. Holli, *Detroit*, 122–23.

50. Fogelson, *Downtown*, 222; Ticknor, "Motor City," 185–87; Detroit Housing Commission, "The Detroit Plan: A Program for Blight Elimination," (Detroit, 1946), from R. Vexler, *Detroit: A Chronological and Documentary History, 1701–1976* (Dobbs Ferry, 1977).

51. Pound, *Detroit*, 309–10.

52. Ibid., 355–58.

53. Ford, "Henry Ford's Home Town," 47–48.

54. T. Munger, *Detroit Today* (Detroit, 1921), 23.

55. WPA, *Michigan*, 232–34.

56. H. M. Nimmo, "Detroit," *American Magazine* 84 (Dec. 1917): 36; Duffus, *Detroit*, 50–59; Munger, *Detroit*, 23.

57. W. Waldron, "Where Is America Going?" *Century* 100 (May 1920): 58–59; Duffus, *Detroit*, 53; E. Wilson, "The Despot of Dearborn," *Scribner's* 90 (July 1931), 24–36.

58. *Outlook* (22 Dec. 1915), 979, 985; Welliver, *Detroit*, 652.

59. Duffus, *Detroit*, 50–51; M. Morse, letter to editor, *New Republic* (31 Aug. 1927), 46; Munger, *Detroit*, 37.

60. H. Whipple, "City Plan Commission Begins Its Great Task of Bringing Order Out of Detroit's Chaos," *Detroit Saturday Night* (3 May 1919), 2.

61. F. Burton, "Record Construction to Continue: Building Head Urges Creation of Central Body to Control Future Problems," *Fordson Independent* (4 June 1926).

62. "Favors Plan for Creation of Metropolitan Area," *Detroit Times* (30 Oct. 1922); "Ford May Help Metropolitan Growth: Appointment of His Engineer to Develop-

ment Committee Called Significant," *Detroit Times* (31 Jan. 1923); Detroit Bureau of Governmental Research, "The Detroit Metropolitan Area" (Detroit, 1924); R. Erley, "The Enlargement of Dearborn," *City Manager Magazine* 8 (July 1926): 9–13; "About the Fourth City," *Just a Minute* no. 16 (15 July 1929).

63. C. Wells, *Proposals for Downtown Detroit* (Washington, D.C., 1942); Reid, *Industrial Decentralization*; E. Wengert, *Financial Problems of the City of Detroit in the Depression* (Detroit, 1939).

Chapter Eleven

Acknowledgments: I would like to thank Robert Lewis and an anonymous reader for helpful comments on earlier drafts.

1. For a survey, see R. Harris, "The Making of American Suburbs, 1900–1950s: A Reconstruction," in *Changing Suburbs: Foundation, Form, and Function*, ed. R. Harris and P. Larkham (London, 1999), 91–110.

2. The closest to a recent exception is B. Nicolaides, *My Blue Heaven: Life and Politics in the Working-Class Suburbs of Los Angeles, 1920–1965* (Chicago, 2002).

3. Small companies that moved factories to the suburbs almost invariably took their office functions and staff with them, and so did many larger concerns. This early, linked decentralization of factory and office remains a neglected issue. See R. Harris and R. Lewis, "The Geography of North American Cities and Suburbs, 1900–1950: A New Synthesis," *Journal of Urban History* 27 (2001): 270–71.

4. J. Vance, "Housing the Worker: The Employment Linkage as a Force in Urban Structure," *Economic Geography* 42 (1966): 294–325. See also S. B. Warner, "If All the World Were Philadelphia: A Scaffolding for Urban History, 1774–1930," *American Historical Review* 74 (1968): 26–43; E. Erickson and W. Yancey, "Work and Residence in Philadelphia," *Journal of Urban History* 5 (1979): 147–82.

5. This implies that where companies established an office and factory at the same suburban location, the requirements of the manufacturing operation were the determining influence. Note that even in the era of the postwar shopping center, retail employment has served, not led, the suburban trend. L. Cohen, *A Consumers' Republic: The Politics of Mass Consumption in Post-war America* (New York, 2003), 257.

6. Citations to chapters in the present volume are not referenced. On Hammond see also J. Bigott, *From Cottage to Bungalow: Houses and the Working Class in Metropolitan Chicago, 1869–1929* (Chicago, 2001).

7. By "alternating" I mean the frequent back-and-forth establishment of homes and workplaces within a specific fringe area. This may produce "leapfrog" development, that is to say, geographically alternating clusters or rings of different land use, or a scattering coupled with progressive infill. Until the widespread adoption of comprehensive zoning in most suburban areas after 1945, the latter was more common.

8. G. Taylor, *Satellite Cities: A Study of Industrial Suburbs* (New York, 1915), 92–97.

9. Ibid., 92.

10. C. Whitzman, "The Dreams Attached to Places: From Suburb, to Slum, to Urban Village in a Toronto Neighbourhood, 1875–2002" (Ph.D. diss., McMaster University, 2003).

11. R. Harris, *Unplanned Suburbs: Toronto's American Tragedy, 1900–1950* (Baltimore, 1996), 82–85.

12. The following is based on evidence reported in J. D. Carroll, "Urban Land Vacancy: A Study of Factors Affecting Residential Building on Improved Vacant Lots in Flint, Michigan" (Institute for Human Adjustment, University of Michigan, Ann Arbor, Michigan, 1952), 16.

13. R. Harris, "Industry and Residence: The Decentralization of New York City, 1900–1940," *Journal of Historical Geography* 19 (1993): 169–90.

14. L. Schnore, "The Separation of Home and Work in Flint, Michigan" (Institute for Human Adjustment, University of Michigan, Ann Arbor, Michigan, 1954), 46. This report contains more local detail than the journal version. See L. Schnore, "The Separation of Home and Work: A Problem for Human Ecology," *Social Forces* 32 (1953–54): 336–43.

15. B. Tableman, "Intra-Community Migration in the Flint Metropolitan District," mimeo (Social Science Research Project, Institute for Human Adjustment, University of Michigan, 1948), 36.

16. Ibid., 41.

17. Schnore, "Separation of Home and Work in Flint," 11.

18. In 1950 the new Chevrolet assembly plant employed the highest proportion of workers who relied on automobiles. Schnore, "Separation of Home and Work."

19. Erickson and Yancey, "Work and Residence," 155; T. Hershberg, D. Light, H. Cox, and R. Greenfield, "The Journey to Work: An Empirical Investigation of Work, Residence, and Transportation, Philadelphia, 1850 and 1880," in *Philadelphia: Work, Space, Family, and Group Experience in the Nineteenth Century*, ed. T. Hershberg (New York, 1981), 138–39.

20. For a survey of Montreal's suburbs, see Walter van Nus, "A Community of Communities: Suburbs in the Development of 'Greater Montreal,'" in *Montreal Metropolis, 1886–1930*, ed. I. Gournay and F. Vanlaethem (Montreal, 1998), 59–76. On Maisonneuve, see P.-A. Linteau, *The Promoters' City: Building the Industrial Town of Maisonneuve, 1883–1918* (Toronto, 1981); on Mont Royal, see L. McCann, "Planning and Building the Corporate Suburb of Mont Royal, 1910–1925," *Planning Perspectives* 11 (1996): 259–301.

21. Reported by Tableman, "Intra-Community Migration," 41.

22. R. Lewis, *Manufacturing Montreal: The Making of an Industrial Landscape, 1850 to 1930* (Baltimore, 2000), 147.

23. See, for example, L. Schnore, "Satellites and Suburbs," *Social Forces* 36 (1957): 121–29.

24. I am assuming here that industrial satellites were the creations of decentralizing industry. Obviously, preexisting satellites that later became centers for decentralizing industry would have been the product of alternating processes.

25. E. Greer, "Monopoly and Competitive Capital in the Making of Gary, Indiana," *Science and Society* 40 (1976): 465–78; R. Mohl and N. Betten, "The Failure of Industrial City Planning: Gary, Indiana, 1906–1910," *Journal of the American Institute of Planners* 38 (1972): 202–15.

26. T. Muller, personal communication, 30 June 2003.

27. See the chapters in this volume by, respectively, Muller, Pudup, and Barrow. On Homestead, see M. Byington, *Homestead: The Households of a Mill Town* (New York, 1910); on the Back of the Yards, see T. Jablonski, *Pride in the Jungle: Community and*

Everyday Life in Back of the Yards, Chicago (Baltimore, 1993); on Hamtramck, see A. Wood, *Hamtramck Then and Now: A Sociological Study of a Polish-American Community* (New Haven, 1955).

28. Schnore, "Separation of Home and Work," 11.

29. L. Odencrantz, *Italian Women in Industry* (New York, 1919) 34; Harris, "Industry and Residence," 176.

30. R. Harris and A. Bloomfield, "The Impact of Industrial Decentralization on the Gendered Journey to work, 1900–1940," *Economic Geography* 73 (1997): 102–6.

31. S. Buder, *Pullman: An Experiment in Industrial Order and Community Planning, 1880–1930* (New York, 1967).

32. Peoria Association of Commerce, *Industrial Resources Survey of Metropolitan Peoria* (Peoria, 1954), 121.

33. "Fastest Growing Town in the Country," *Peoria Journal-Transcript* (13 Dec. 1936), sec. 4, 1.

34. L. Carr and J. Stermer, *Willow Run: A Study of Industrialization and Cultural Inadequacy* (New York, 1952).

35. S. B. Warner, *Streetcar Suburbs: The Process of Growth in Boston, 1870–1900* (Cambridge, 1962).

36. E. Taaffe, B. Garner, and M. Yeates, *The Peripheral Journey to Work: A Geographic Consideration* (Evanston, 1963), 13.

37. There is an extensive literature on the impact of the automobile. For a concise assessment, viewed in historical perspective, see J. Vance, "Human Mobility and the Shaping of Cities," in *Our Changing Cities*, ed. J. Hart (Baltimore, 1991), 67–85. Vance's earlier research shows an exemplary appreciation for the changing nature of the home-work linkage, one that has rarely been equaled. See J. Vance, "Labor-shed, Employment Field, and Dynamic Analysis in Urban Geography," *Economic Geography* 36 (1960): 189–220; and Vance, "Housing the Worker."

38. G. Smerk, "The Streetcar Shaper of American Cities," *Traffic Quarterly* 21 (1967): 578–79.

39. Harvard University Graduate School of Design, *Framingham: Your Town, Your Problem* (Cambridge, 1948), 21.

40. H. Conant, "The Locational Influence of Place of Work on Place of Residence" (master's thesis, University of Chicago, 1952).

41. J. Carroll, "Some Aspects of the Home-Work Relationship of Industrial Workers," *Land Economics* 25 (1949): 418.

42. Ibid. See also A. Hawley, "The Willow Run Area," mimeo, report by the Willow Run Community Council, 30 June 1943.

43. There is an extensive, dispersed literature on the Levittowns. See, notably, B. Kelly, *Expanding the Dream: Building and Rebuilding Levittown* (Albany, 1993).

44. Vance, "Labor-shed," 214.

45. J. Sewell, *The Shape of the City: Toronto Struggles with Modern Planning* (Toronto, 1993).

46. R. Fishman, *Bourgeois Utopias: The Rise and Fall of Suburbia* (New York, 1987).

47. Nicolaides, *My Blue Heaven*.

48. Ibid., 75–78. Even these figures are probably overestimates. No data were available for a number of companies, including some of the largest in the area. Judging from

the available evidence reported by Nicolaides, the employees of the larger companies were most likely to commute from farther afield.

49. For a useful discussion of sources that overlooks directories, see L. Schnore, "Three Sources of Data on Commuting: Problems and Possibilities," *Journal of the American Statistical Association* 55 (1960): 8–22.

50. A. Bloomfield and R. Harris, "The Journey to Work: A Historical Methodology," *Historical Methods* 30 (1997): 97–109.

51. See also I. Katznelson, *City Trenches: Urban Politics and the Patterning of Class in the United States* (New York, 1981).

About the Contributors

Heather B. Barrow is a Ph.D. candidate in the history department at the University of Chicago. Her dissertation examines the formation of Dearborn within the context of Detroit's interwar metropolitanization.

Gunter Gad is professor of geography at the University of Toronto. He has written extensively on the formation of the North American central business district and the historical geography of Toronto's office and manufacturing districts.

Paul A. Groves is deceased. He was an associate professor of geography at the University of Maryland, College Park, and the author of articles on the historical geography of North American cities, with special emphasis on Washington, D.C. He co-edited the first edition of *North America: The Historical Geography of a Changing Continent*.

Richard Harris teaches urban historical geography at McMaster University. He has written about residential segregation, home ownership, the building industry, and suburban development in North America. His history of Canadian suburbs, *Creeping Conformity: How Canada Became Suburban, 1900–1960*, appeared in 2004. He is currently doing research on the emergence and evolution of housing policy in the British colonial territories from 1929 through independence.

Greg Hise is associate professor in the School of Policy, Planning and Development at the University of Southern California. His research examines American cities and regions since 1850, with particular attention to southern California and the American West. His books include *Eden by Design: The 1930 Olmsted-Bartholomew Plan for the Los Angeles Region* (2000), with William Deverell, and *Magnetic Los Angeles: Planning the Twentieth-Century Metropolis* (1997). His current research examines the landscape history of greater Los Angeles from 1850 to 1930 and the environmental history of southern California.

Robert Lewis is associate professor of geography at the University of Toronto. His recent book, *Manufacturing Montreal: The Making of an Industrial Landscape, 1850 to 1930* (2000), examines the formation of Montreal manufacturing districts. He is currently writing a book on metropolitan Chicago, 1865–1940, with particular attention to inter-firm networks.

Edward K. Muller is professor of history and director of the Urban Studies Program at the University of Pittsburgh. His research has recently focused on urban planning in Pittsburgh as well as the formation of metropolitan regions. He is currently writing three books: a history of planning in Pittsburgh between 1889 and 1943 (with John Bauman), a popular history of Pittsburgh (with Rob Ruck), and an edited collection of Bernard DeVoto essays on conservation and the American West.

Mary Beth Pudup was trained as a historical geographer of regional economic development in the United States. She is currently associate professor of Community Studies at the University of California, Santa Cruz, where she teaches classes on political economy, urban studies, and social change activism.

Richard Walker, professor of geography at the University of California, Berkeley, has written on a diverse range of topics in economic and urban geography, as well as on environmental policy, philosophy, and California studies. He is the author of *The Capitalist Imperative: Territory, Technology, and Industrial Growth* (1989), with Michael Storper, and *The New Social Economy: Reworking the Division of Labor* (1992), with Andrew Sayer. He is currently at work on a book on the economic, political, and cultural geography of the San Francisco Bay area.

Index

Individual suburban and industrial areas can be found under the relevant metropolitan district.

189, 191–97; southern California, 178, 181, 188, 190, 192; South Gate, 234–35; surveys of, 182, 185–86, 192; Torrance, 178, 182, 186, 188–89, 195, 198; Vernon, 29, 179, 186, 189, 192, 193, 195, 197, 199
Los Angeles River, 193, 199
Los Angeles Soap Company, 184–85
Lowell, 29, 37

manufacturing district. *See* industrial districts
Martin, Glenn L., 197–98
Massey-Harris (Massey Manufacturing) Company, 150, 163–64, 173
McCormick Bros. Reaper Works, 63, 67
McKenzie, Roderick D., 142
meat packing industry, 227; in Chicago, 64, 69, 70–73; in San Francisco, 102, 104–6, 108, 110, 111, 112; in Toronto, 153, 154, 173, 174, 176. *See also* stockyards; Swift and Co.
Mellon family, 132, 133, 135, 141. *See also* growth coalitions
merger movement, 150. *See also* capital
mesogeography of urbanization, 26–27
metal-working industry, 8, 158, 188; and business linkages, 66, 136, 139; geography of, 37, 98, 100, 101, 173, 183; and industrial growth, 38, 44, 51, 97; machinery, 24, 62, 115, 130, 154, 169, 173, 185, 188; machine shop, 40, 49, 110, 115, 139; as producer goods, 36; and suburbanization, 12, 81, 86, 106, 115, 153, 154; and workers, 8, 83, 187. *See also* foundries
metropolitan area, 54, 179, 202; dispersed industrial clusters, 186; and growth coalitions, 87; intra-metropolitan competition, 181, 182; multinucleated, 22–25, 54, 74, 78, 79, 86, 174, 202; metropolitanization in Detroit, 200–203, 219, 220; in Pittsburgh, 124–25, 142; and production, 9–10; and spatial scales, 10, 24; and suburbs, 12–14; and urban form, 17, 24, 79–80. *See also* suburbanization
military reserve lands, 156
Miller and Lux, 101, 102, 104–6
mill towns, 35
Milwaukee, 2, 70
mining, 10; in California, 93; coal, coke, and gas, 10, 130, 133–36; Connellsville coke district, 134–37; natural resources, 10, 12, 55, 68, 93, 125, 133, 141; patch towns, 142
Monongahela River, 12, 126, 127, 129, 130, 131, 133, 135, 137
monopoly capitalism, 146. *See also* capital

Montreal, 9, 76–91, 151; Canal, 81, 84, 85, 86, 90; and capitalist industrial suburbanization, 78–79; Hochelaga, 83, 85, 90; industrial district, 24; industrial suburbanization, 7, 8, 12, 79–85; Lachine, 12, 83, 86, 87, 89, 90, 225, 226; Maisonneuve, 6, 29, 86–87, 89, 90, 226; Mercier, 88; metropolitan suburbanization, 85–91; Montréal Est, 87, 88, 90; Montreal Harbour Commission, 85; Mont Royal, 225; North End, 90, 91; rivalry with Toronto, 150; Rosemont, 88; St. Henri, 83, 85, 86, 87, 89, 90; Ste. Cunegonde, 83, 85, 89; Ste. Marie, 83, 85, 86, 87, 90; Verdun, 87; Ville St. Pierre, 86
Munger, Thomas, 216–17
municipal, 155; boundaries, 10–12, 28, 45, 144, 156, 158, 166, 167, 175, 189, 204–26, 224; industrial policy, 151, 152–53, 162, 180, 182, 190–92; management, 205; taxes, 151, 152, 205, 206, 207, 208, 209, 224

Natick, Mass., 233
New Orleans, 37
New York City, 108, 183, 185, 189, 201, 215; and metropolitan growth, 4, 13, 192, 224, 228–29; office space, 214; transportation, 151, 227, 230
Nimmo, H. M., 217
noxious industries, 2, 34–35, 41, 101, 182, 186
nuisance ordinances, 74

Oakland, 22, 104, 107, 109, 111–19, 122, 224; early industrial districts, 111–12; housing and residential areas, 112, 116–17; industrial and population growth, 111, 113–16; land development, 112–13, 117; politics of space, 117–19; relations with San Francisco, 117–19. *See also* San Francisco
offices, 213–14, 224, 231
Ohio River, 126, 127, 131, 137
Ohio Valley, 37, 38
Olmstead, Frederick Law, 132
Olmstead, Frederick Law, Jr., 188
organized industrial districts. *See* planned industrial districts

Paterson, N.J., 4, 12
Peoria, 229–30; East Peoria, 229
petroleum refining, 86, 87, 110, 121, 196–97
petty commodity production. *See* production organization